Transfers of Manufacturing Technology Within Multinational Enterprises

Transfers of Manufacturing Technology Within Multinational Enterprises

Jack N. Behrman
Harvey W. Wallender

This study is the result of a project sponsored and administered by the Fund for Multinational Management Education

Ballinger Publishing Company • Cambridge, Massachusetts
A Subsidiary of J.B. Lippincott Company

T
174.3
B43

 This book is printed on recycled paper.

International Standard Book Number: 0-88410-048-0

Library of Congress Catalog Card Number: 76-5866

Printed in the United States of America   3 3001 00594 3849

Library of Congress Cataloging in Publication Data

Behrman, Jack N
    Transfers of manufacturing technology within multinational enterprises.

    1. Technology transfer. 2. International business enterprises.
I. Wallender, Harvey W., joint author. II. Title.
T174.3.B43   338.4'5   76-5866
ISBN 0-88410-048-0

Contents

298670

List of Figures

List of Tables

Preface

Much has been written and argued about the transfer of technology over the past fifteen years in an effort to formulate better governmental policies for promoting adoption of technologies. But little effort has been made to describe the process in detail. This study was conceived as a step toward filling that void; hopefully, policies could then be based on a more adequate understanding of the complex processes at work. It soon became evident, however, that it is impossible to describe fully the transfer of technology among members of a multinational enterprise. The process is one of continuous communication, and only an hourly replay could provide a detailed description. The cases given here, therefore, are partial and incomplete but hopefully sufficiently detailed and accurate so that the outlines of the process are correctly displayed. No more than that was hoped for or sought, and no assessment has been made of whether the companies described were performing well or badly, whether deficiencies or gaps in the technologies transferred existed, or whether governmental policies fostered or circumscribed such transfers in ways that were undesirable from the standpoint of either the company or national economic policy.

Any lessons will have to be drawn by the reader himself, though there are some fairly clear conclusions that fall out of the cases. However, more studies should be made, and more companies should tell their stories openly.

The selection of these particular case studies was made with the objective of obtaining a significant cross-section of multinational enterprises, both industry-wise and geographically. Several companies

were approached with the idea of telling the story of their technology transfers, and discussions occurred over several months as to
what would be required, the methodology, the time frame, scope,
and so forth. Only one company among those approached found that
it could not cooperate, though some of its officials thought that
they had an interesting story that should be included.

We sought companies having affiliates in several different regions,
so that any lessons drawn would not be warped by too narrow a base
of description. Readers can make some comparisons among affiliates
of the same company, companies employing quite different technologies, and affiliates of quite different size and experience in
receiving technology. After the companies agreed to cooperate, the
selection of the precise affiliates to be studied had to be agreed upon.
The technologies were not to be too simple, yet could not be so
sophisticated that nonexperts would be unable to comprehend and
describe them adequately. And different countries were sought so
as to illuminate differences in transfer processes caused by abilities
of affiliates to receive technology.

It was decided to focus solely on transfers among members of
multinational enterprises (i.e., affiliates wholly- or predominantly-
owned by the parent) so as to remove all questions of constraint of
transfers arising from ownership patterns. These cases, therefore,
describe what has occurred in "family" situations. This facet of the
study design was prompted partly by the wide focus in governmental
policies on independent licensing of technology, and by the frequent
proposals of experts that technology be purchased in the market,
apart from ownership relations. Much more U.S. technology is
probably transferred to wholly-owned affiliates than to independent
licensees, given the number of affiliates compared to licensees and
the scope of technologies transferred to each. Yet, governments
seem more interested in the independent licensing; for example, the
regulations on technology transfers recently promulgated by Mexico,
Argentina, Venezuela and the Andean countries, and by the ASEAN
countries are oriented toward independent licensing. In addition, the
codes proposed by the Pugwash Conference and by the International
Chamber of Commerce relate largely to transfers outside of wholly-
owned affiliates. If independent licensing is to be proposed, and
joint ventures are urged, some comparison of the resulting technology transfers should be made; one base for comparison would be
the process of transfers under wholly-owned situations.

An examination of these cases should throw some light on policies
other than those toward technology transfers—such as those on
product changes, ownership requirements, control and decision

making, technical training institutes, and promotion of local R&D facilities in host countries—though none of these were directly studied within these cases.

In order to focus the study sharply, it was decided to concentrate on the transfer of manufacturing technology only and to exclude the transfer of management, marketing, and finance skills, despite the fact that all of these are closely related to successful transfer of technical know-how. It would be difficult enough, we thought, to try to encompass the whole range of technical transfers related to manufacturing operations—from the conceptualization of the project, site selection, construction, start-up, continuing operations, to product changes.

Not only would we be concerned with describing the full range of technologies transferred but we also would have to pursue the process throughout the company—from parent to affiliate, from parent to management center to affiliate, from affiliate to affiliate, and from the final recipient back up the line in the form of adaptations and improvements. The description would not be complete unless it also moved outside of the company into suppliers (whether local- or foreign-owned), into customers and repair services, into competitors or other companies through personnel transfers, and into worker lifestyles.

Yet, as indicated above, it was recognized that it would not be possible to tell the "whole" story. It would not be possible in a book this size even to reproduce the tables of contents of the manuals and specifications that were transferred to the several affiliates over their lives by the parent companies. In one instance, the mere construction manuals on which the manufacturing manager relied constituted fourteen volumes of two-inch thick looseleaf binders. We had anticipated, however, that we could "tell an adequate story" within about half of the space that we have finally consumed in this volume. The "story," therefore, is both longer and more inadequate than we had anticipated at the beginning—which merely shows that we also have gone through a substantial learning process.

Since there are few written reports, even by the companies themselves, on the process of technology transfer among company units, it was not feasible to gather company documents and sit down to write a "research piece." We sought, therefore, permission to spend several days with company officials around the world discussing procedures and problems in the transfer process.

Interviews were held with officials of the parent company and visits made to the plants of the parent company that were backstopping the affiliates abroad, in order to see the process of manufacture and to become familiar with the vocabulary used. Visits were then made to the regional management centers to talk with the officials in charge of technology transfers and product development. Each of these visits required a day or more of conversations. A week's interview was planned for each affiliate studied, and between four and five days were actually required at each to talk with all relevant officials, gather documents, and visit the plants. At such visits, it was frequently also possible to talk with supplier companies or customers to discuss the scope of technologies received by the local affiliates. And special efforts were made to talk with some of the workers about their training, promotion within the company based on technical advancement, and the extent to which they used technical skills learned there outside of the company.

The persons interviewed were those directly responsible for and affected by the transfers, beginning with the top manager, his manufacturing chiefs, technical staff officers, plant managers, quality assurance personnel, supervisers and foremen, trainers and workers. In addition, we obtained (or asked to be collated, when they were not gathered together already) records of visits from parent-company officials and management centers to affiliates, among affiliates, and from affiliates to parents. We examined specifications transferred, report forms, policy statements on technology transfers, agenda for regional meetings of technical personnel, training programs, inspection procedures, and organization charts.

The results of the interviews were combined with the information gathered from the documentation and a draft written. This draft was then redistributed to the parent, management center, and affiliate for checking for accuracy and omissions. The technical nature of the transfers meant that nontechnical researchers could well misunderstand information or processes. Some few errors were caught in this process, but the substance of the reports was not altered. Rather, the companies found their own story so interesting that they contributed additional information to make it more complete, and some officials who were initially somewhat reluctant about the study have urged that more such stories be researched for affiliates of the company so as to instruct their own officials about what is occurring within the company. Some officials were quite intrigued by the amount and type of information that was developed in the cases, and no significant deletions were requested—only the names of individuals so as to avoid embarrassment to those involved in particular problems, those who may have left the company, or those outside the company.

We obviously could not have produced these company histories without the full cooperation of many officials in each of the locations visited. Since top management had given approval, one would expect cooperation at each division or affiliate, but we were received with such fulsome cooperation that there is little doubt in our minds that every facet of significance in the transfer of technology has been at least touched upon. In some instances, the company officials answered questions on facets that they would not have included in "technology transfer" but which they later recognized as an integral part thereof.

It is not possible, nor appropriate, for us to list all those who were responsible for our being able to recount these exchanges of technology. We do wish to express here our sincere appreciation for their making the task a most interesting and enjoyable one for us both.

Transfers of Manufacturing Technology Within Multinational Enterprises

✳ *Chapter 1*

Introduction and Conclusions

In its report on "The Impact of Multinational Corporations on Development and on International Relations" The Group of Eminent Persons of the United Nations stated: "The multinational corporations have become the most important sources of a certain type of technology. Their affiliates can draw upon the knowledge of the entire organization of which they are a part . . . . in practice, however, the full transfer of knowledge may not take place; partly because it is not always suitable for use by the affiliate and partly because the parent company will not always wish to make it available."[a] The report then goes on to argue that developing countries, particularly, need to be able to buy the technology they need and they have to have much more information about what technologies are useful and available in order to be able to buy effectively. The UN Group also questioned the particular kinds of technology transferred and the products with which it is associated.

Much of the concern of the UN Group was with the ability of the developing countries to purchase or acquire specific technology. Though the Group recognized that there was a more or less automatic flow of technology from the parent to an affiliate in a wholly-owned relationship, most of the discussion centers on joint venture or independent licensing arrangements. In the wholly-owned situation, the initiative is with the parent to a significant degree, whereas under the minority ownership or licensing arrangements the initiative is largely with the recipient. Therefore, there is a bias in the

[a]U.N. Department of Economic and Social Affairs, Document E/5500/Rev. 1 New York, 1974, page 66.

orientation of the Group towards less than wholly-owned affiliations between the subsidiary and the parent. This bias reflects the increasing desire on the part of developing countries for control and for selection over the products produced and the technologies employed.

However, as the quotation above indicates, a wholly-owned affiliate has at its finger tips the entire technical know-how of the parent and, as shown in the studies in the subsequent chapters, the experience and knowledge of other affiliates as well. This range of technical knowledge is not available to minority-owned affiliates or to independent licensees. Therefore, the flows of technology are quite different in these three major relationships: wholly-owned affiliates, minority-owned, and independent licensing. The selection of the latter two routes may occur at very high cost in terms of technical assistance from the international companies.

It was for the purpose of getting a better picture of the complex flows of technology from the parent to affiliates and among affiliates that we chose in this series of studies to focus on wholly-owned (or predominantly-owned) affiliates.

After one examines the experience under these associations, the specific recommendations of the Group of Eminent Persons seem to be highly limited in application. The several recommendations on technology were as follows:

The Group recommends that before a multinational corporation is permitted to introduce a particular product into the domestic market, the host government should carefully evaluate its suitability for meeting local needs.

The Group therefore recommends that the machinery for screening and handling investment proposals by multinational corporations, recommended earlier, should also be responsible for evaluating the appropriateness of the technology, and that its capacity to do so should, where advisable, be strengthened by the provision of information and advisory services by international institutions.

The Group recommends that host countries should require multinational corporations to make a reasonable contribution towards product and process innovation, of the kind most suited to national or regional needs, and should further encourage them to undertake such research through their affiliates. These affiliates should also be permitted to export their technology to other parts of the organization at appropriate prices.

The Group draws attention to the work of the Economic and Social Council and UNCTAD on technology (including decision 104 (XIII) of the Trade and Development Board on exploring the possibility of establishing a code of conduct for the transfer of technology) and recommends that international organizations should engage in an effort to revise the patent system and to evolve an over-all regime under which the cost of technology provided by multinational corporations to developing countries could be reduced.

The Group recommends that host countries should explore alternative ways of importing technology other than by foreign direct investment, and should acquire the capacity to determine which technology would best suit their needs. *It also recommends* that international agencies should help them in this task.[b]

As can be seen, these recommendations aim quite specifically at the less than wholly-owned relationships, though they also would apply to them. Since, as the members of the Group of Eminent Persons state, decisions on policy matters such as these should be derived from substantial information on the impacts of the policies and the decision criteria of international companies, it is of critical importance that the entire story of technology transfers be told.

Despite extensive studies on technology transfers, the present level of research and discussion on the relationship of technology transfers and development has failed to provide adequate policy guidance for evaluating and controlling the full range of technologies required in developing the different manufacturing sectors. Part of the inadequacy of the dialogue results from the intensity of political feelings surrounding foreign investment and technology, but part of it is the result of repeated narrow definitions of technology, which separate the elements of technology transfers that seldom in business practice are as readily separated as the theorists imply. Thus, technology under patents is separated from technology that would have to go with the patent to make a knowledge there understandable; manufacturing technology is separated from design technology; product engineering is separated from testing technology; and production engineering is separated from plant design and construction—all as though one could buy a piece of a total system and make a different system work as effectively. This is not to say that pieces of technology cannot be useful, but to understand technology transfers requires an understanding of the environments out of which the technology is supplied and into which it is placed.

This separation of technologies on the part of researchers and theorists is mirrored in public policy guidelines, where regulatory and advisory bodies dedicate their efforts to specific technologies rather than to the overall transfer process. This lack of attention to the total process is a signal failure on the part of both governmental groups and company managers who are responsible for it.

[b]Ibid., pp. 68–70.

This separation of technologies in the transfer process was supported further by the development of licensing of patents and particular industrial know how as occurred in the 1950s and 1960s, the use of technical-assistance agreements, and the emergence of turn-key operations, which separated the total technical package from continuing management by the supplier of the technology.

Contrary to the approach of public research, which focused on specific elements of technology transactions, the decision making of the companies aggregated the technology transfers into total business activities, as one element of total indices of profitability or productivity. Thus, the government analyst and the supplying firm were looking at the situation from different viewpoints. The international companies had not segregated their concept of technology from the entire business picture, and the government analyst could not relate separate technology elements to the overall business venture.

From the case studies in the subsequent chapters, it is possible to develop a structure of technology transfers, which should help each party understand the process better and therefore appreciate the viewpoint of the other.

TYPES OF MANUFACTURING TECHNOLOGIES

Technology transfers in most manufacturing activities occur over seven distinct phases: three prior to start-up of the plant and four subsequently. The three early phases include (1) initiation of proposals for site location and planning of the operation, (2) product designs, and (3) design and construction of the facilities. Once the plant facility is completed, industrial engineering and training plays an important role in start-up activities and often requires a substantial number of personnel from the parent. Subsequently, assistance is provided in the value engineering phase, which involves the establishment of controls and testing procedures to maintain quality and assure the customer of appropriate product standards. A later phase relates to the introduction of new products and product development. The final type of technology transfer brackets almost all of these phases since it relates to technology support provided local suppliers in order to improve their products, to develop new sources of labor, to help establish industrial standards or specifications in the host markets, and to instruct potential customers on the varied and specific uses of the product; these may be called external-support transfers of technology.

For each type of technology, different transfer mechanisms will be found appropriate. Much of the specialized technological content at the early stages will likely come in the form of architectural drawings, oral advice by visiting engineers or industrial architects, and instruction on use of special equipment. Five general mechanisms of transfer are evident: (1) documentation in the form of manuals, specifications, layouts, designs, and so forth, which may be produced for specific assignments or through regular reporting procedures among affiliates or from the parent; (2) instruction programs—that is, formal education, and on-the-job training (OJT), including intracompany seminars and conferences among technicians and experts; (3) visits and exchanges of technical personnel; (4) development and transfer of specialized equipment; (5) and continuing oral and written communication on whatever problems may arise at the affiliate.

The relationship of these types of technology and the mechanisms available to transfer them can be conceptualized in a matrix form, as shown in Table 1-1, which demonstrates the complexity of technology transfers and the ways in which they may be analyzed though not separated. Segmentation, in fact, destroys one of the major advantages of technical ties—that is, the ability to tap the entire technical knowledge of the parent and the affiliates of an international company.

The matrix in Table 1-1 can be of considerable use to companies in understanding what they are transferring, how it is transferred, to whom it goes, and how it is used. It is of equal value to government agencies seeking to determine what kinds of technology they want and what the costs are of precluding transfers of specific types of technology. Such an assessment shows fairly quickly that it is virtually impossible to cost-out these transfers of technology. Not only is it virtually impossible to cost-out separate pieces of the technology, but also it is impossible to determine the cost of the entire technology package. This conclusion will be more readily understood as one examines the case studies in the subsequent chapters.

From these studies there emerges a fairly precise picture of the structure of technology transfers, which is detailed in the following subsections.

Manufacturing Proposal and Planning

Technical assistance to any new manufacturing project begins in the very first stages of a proposal. At this point, specialized personnel within the parent company are drawn together to help determine the product line, the scope of the operation, site location,

Table 1-1.   Technology Transfer Matrix

Mechanisms / Types	Documentation		Instruction		Conferences and Seminars	Visits and Exchanges	Equipment	Communication on Problems
	Manuals, Spec'ns, Designs, Drawings	Regular Reports	Formal	On-the-Job Training				
Proposal & Planning								
Product Design								
Plant Design & Constr'n.								
Start-up								
Value Engineering and Controls								
Product Development								
External-support								

production techniques, and budgetary aspects of the manufacturing operation. These personnel will have drawn on their experience not only within the parent, but in establishing other such facilities around the world. In addition, they will do preliminary studies of the possibilities of employing host-country suppliers, technicians, and different types of skilled labor.

To obtain the necessary information in formulating production plans, teams of experts and organization specialists will normally be dispatched to the new site to work with locals in developing detailed plans and guidelines for plant construction and operation. These specialists will draw on company manuals, their past experience, corporate cost-accounting guides on each of the phases with which they are concerned, evaluation models developed by the company, and a variety of technical documentation that will limit the necessity to go through duplicative investigations and studies. A recipient country that purchases a turn-key operation misses out almost completely on this phase of technical assistance—at least in terms of learning what is being done. Though the results of this exercise are incorporated in the purchased facility, there is little transfer of technology through the educational process. For a wholly-owned operation, however, those who will carry on the manufacturing operation are frequently drawn from these early project teams or are brought into those teams from the host country for the purpose of learning the entire process of manufacturing development.

Similarly, in straight sales of technology, the recipients have little information regarding how the supplier views the initial problems of use of manufacturing technology and why it recommends the ultimate techniques or machinery and equipment that are transferred. If the receiver were carried through the initial proposal and planning stages, he might be more willing to sacrifice certain elements of the technology package to gain others, in order to adapt more effectively the technology to his own needs and objectives. Where technology agreements are derived from bargaining situations, rather than out of wholly-owned relationships, the competitive position of the supplier and receiver often preclude any cooperation at the preliminary planning stages. Consequently, the short-term focus on "low cost" and a "good deal" in economic terms may be short-changing both parties in terms of the total technology transferred for the long term success of the project.

To grasp the scope of assistance that occurs during this stage, it is only necessary to briefly characterize the information, guides, and decisions required in this phase:

Site selection and environmental analysis;
Operating estimates;
Preliminary design analysis—product design, production layout, including maintenance and repair facilities, power plant, and so forth;
Cost calculations—plant and operation;
Organization memoranda—project management, division management, staff support, and so forth;
Personnel selection and qualification;
Planning and control documents—problems expected and their treatment.

If the recipient obtains answers to these problems merely in the form of a new plant without participating in the decisions leading to the particular operations, he is less well prepared to handle the problems when they arise. The skills required to develop the policies and guides at this stage should be transferred in some degree to the management of the new facility in order to insure continued growth and flexibility. Yet, the techniques and procedures needed to develop a proposal—make preliminary plans and analysis—are often the hardest to acquire and the most important for the long-term growth of the company.

Product Design

Unless there is a direct carry-over of a product or product line of the parent into the foreign affiliate, some adaptations in product design will be necessary for the foreign market. These adaptations may be so extensive as to produce a virtually new product, or they may merely be adaptations necessary to adjust to different inputs of materials or qualities of air and water.

Few facilities in developing countries have the capability of making such adaptations, and unless they are part of an international company, they will not obtain the technology necessary to make such adaptations—that is, they will be given an adapted product, without even knowing necessarily why the adaptations were made or what problems they might lead to.

Product design starts, of course, with some concept of the market that is gleaned either from a market analysis or market experience in other countries. From there, it goes into a series of elements making up the total design package:

Styling;
Mechanical attributes;

Standardized componentry, compatibility with existing peripherals;
Power—hydraulic, electrical, mechanical—materials inputs;
Packaging.

Frequently, the skills needed to design a product are simply not purchasable in the open market—that is, it would be possible to buy the designs for an existing automobile and produce it in a developing country, but it would be quite difficult to obtain in the market the technicians necessary to produce a new automobile, designed from scratch. This takes a degree of expertise held only within existing companies, who are not likely to lease their expertise or technicians to an outside company.

Plant Design and Construction
In the process of designing a plant and carrying out the construction, a significant volume of technology will be transferred to the recipient country—either in a turn-key operation or through local technicians, such as industrial architects, construction engineers, contractors, specialists in air conditioning, water treatment, power train, and so forth, who are hired for the project. The international company would bring in engineers, architects, and designers to provide detailed layouts, machinery plans, architectural drawings, and assistance as well as other basic information in plant design and construction. Besides these technical aspects, much assistance is given in the management of construction itself. This is a form of technical assistance that is frequently very much needed in the developing countries.

During this phase, teams of experts are sent to the site area from the parent, while locals who are working on the project and are intended to be kept with the company are exchanged back to headquarters to gain specialized information on the design and layouts of the facility so that they can maintain, repair, and alter as needed in the future. The specialized skills that have to be developed include the following:

Equipment lists and layouts;
Site work designs;
Air conditioning and water requirements;
Transport and power specifications;
Laboratory and plant equipment specifications;
Modification of standard designs to local conditioning.

Construction management which includes the necessity to select and supervise bids from local contractors, suppliers, and architects, is a separate and distinct skill, though one that can be learned through technology transfers. Where an international company is setting up the operation, it also has the facility of coordinating local designs with its global purchasing plans and creating standard elements of the design and construction, which a local enterprise would not be able to do—at least as well.

Because of their long experience in many areas, a number of international companies have specialized field and design engineers who are formed into project teams to work on this particular phase in several countries. These specialized technicians will likely have participated in design and construction of many other plants around the world. Their collective capabilities are normally hard to secure and harder to take advantage of, unless locals are encouraged to work closely with them, thus creating specialized "on-the-job training" during the design and construction phase.

Start-Up Phase

Preparation for the start-up phase begins with the transfer of production technology, which includes written procedures, processes, specifications, manuals, quality-control guides, and a wide variety of other documentation. Specialists are sent over extended periods to install machinery, make appropriate tests, and establish quality-control procedures. Specialists are also sent to begin extensive training at the plant as well as in formal classrooms; and some of the future engineering specialists and managers are returned to the headquarters company for training. Permanent personnel are, therefore, simultaneously trained on-the-site and at corporate headquarters or at other affiliates that have similar operations. Provision must be made for (and future managers educated on) repair and maintenance systems, procurement procedures, manpower development programs, production scheduling, and a wide variety of other engineering management and production procedures. At this stage, the roles of the production and industrial engineers are most important and are given primary attention by the parent company.

To get an idea of the range of technologies transferred, several types of activities must be carried out simultaneously, before initial production can begin:

Procedures for purchase of supplies;
Cost-control guides and routines;
Direct labor hired and trained;
Indirect labor hired and trained;
Quality-control tests and procedures;
Production routines;

Maintenance and engineering facilities;
Equipment tested and adjusted;
Test equipment installed and tested.

In the establishment of a new facility, this is the phase at which most production technology is transferred. All of the elements and materials handling, packaging, production, and quality control must be transferred and be successfully adopted in this phase. Therefore, manpower development for the entire plant must have been completed, foremen and supervisors selected and trained, and production and industrial managers and engineers selected and trained—a process which can require from one to three years on and off site.

These individuals must have been supplied and have learned production manuals, test procedures, production routines, test routines, and so forth in order to build a solid foundation for dealing with production difficulties and scheduling. To assist them, generally, a specialized manufacturing start-up team is provided for several days or weeks of initial operation. Before they leave, the entire technology for the basic plant should be operational. In addition, in order to replace quickly the natural turnover of both direct and indirect labor in the new facility, an internal training capability must be provided; for example, for a three-hundred worker plant with turnover of 15 percent at all levels, a training facility providing for between 1,500 and 4,000 man-days of training annually will be required.

After a few weeks or months of operation, even a turn-key contract brings the cessation of further technology transfers. What will have been learned is essentially only this last phase of start-up, when the foreign technicians are turning over the operations to locals. But, technology has a dynamic quality and is sequential in nature. By the time the plant is constructed and operating, there will already be new techniques and equipment that could be used to improve the efficiency of operation or assist in adapting to the new competition of product requirements. In fact, the international company will have learned new ideas and techniques in the construction of this very plant—lessons that the recipient will not have learned because it did not have the same base to begin with. These new lessons will be used by the international company in the next operation in another country, while the present recipient will simply not be well informed. However, if this is an affiliate on an international company, so that there is a continuing flow, whatever is needed at the affiliate will be available to it.

In addition, a continuing relationship provides a better understanding of how to communicate across cultural barriers. Not only will the international company take more time and instruction and care in communicating the ideas, but also it will spend some time in trying to learn the problems of cross-culture exchanges. There is not the same pressure to take this care when the transfer is under a turn-key operation or under licensing to an independent company. Also, the transfers of technology are likely to be more effective, because the donor company knows more thoroughly and precisely what the recipient requires in engineering, processes, quality control, equipment, performance criteria and so on, for he has planned the operation or participated fully and it reflects much of what exists either at the parent company or another affiliate. Thus, there is a quick familiarity with the problems that are likely to occur and a much more ready solution is found, than if the recipient's operations are strange—as they frequently are to a licensor who must then take time to familiarize himself with the particular set-up at the licensee.

Value Engineering

Once the plant has adjusted the new equipment and processes, the focus of attention is shifted toward improving its cost-effectiveness. A great deal of time of the engineers and managers is spent in applying techniques and procedures to improve productivity and raise quality. Inspection procedures, control procedures and techniques, tests—all are re-evaluated and modified to maximize the plant's activities in relation to its local environment. At this stage, the plant may rely on developing its own innovating techniques transferred from other affiliates. Much of progress in value-engineering comes from an exchange of ideas among different production and engineering teams within the plant and among affiliates. Most international corporations encourage these exchanges through seminars and programs designed to create not only a professional competitive relationship but a free exchange of information on ideas worldwide, especially among affiliates they control. This free flow is facilitated by rapid communication. The telex often represents a vital link for questioning and problem-solving. A request for information on supplies or a specific production bottleneck can be fed to headquarters where it is analyzed and then passed on to plants with similar situations. Thus, the requesting facility receives the benefits of multiple experiences. Professional seminars are another way of formalizing the human contact that stimulates new ideas and techniques. Some companies require all specialists and managers to spend up to thirty work days annually on exchanges or seminars. Naturally, new information is routinely distributed through revised manufacturing instructions and memoranda on plant problems.

Specifications and recommendations for equipment are also routinely available to all the plants.

When the technology planner or analyst focuses too narrowly on specific technical know-how, he overlooks the importance of on-going technology links and stimuli for improvement and adaptation of processes. The reports of plants and sites constructed and later idled due to a lack of on-going technical support are numerous. While in many developing countries, national planners secure the best equipment and production systems, they fail to provide for on-going support and contact for the facility with a source of continuing know-how, so that it either withers or becomes grossly uncompetitive and thus requires protection and has no opportunity to export.

Product Development

If an effective technology base has been established, the new plant and its support activities should be capable not only of improving its basic procedures but also of developing new techniques and products.

A manufacturing facility will normally evolve an independent product development and industrial-engineering capability sufficient to sustain itself in most areas. However, transferring product-development capability is much more difficult. Much of the basic research and technology core must be centralized to allow the economies of scale and concentration of expertise necessary for breakthrough research and effective utilization of highly specialized skills. No affiliate could justify having the variety of expertise needed to sustain development of the wide range of new products needed to stay competitive.

The transfer of developmental facilities is also difficult because of barriers raised against the importation of equipment and supplies and their lack of availability locally; speciality chemicals, laboratory animals, or electronic data processing are examples. Finally, laboratories capable of developmental research demand well-qualified technicians and scientists who often are not available for industrial research.

At this moment, most companies can effectively transfer the capability of development of some independent products, but it is not always in the interest of the affiliate or host country to do so—especially if such differentiation precludes participation in the world market.

External Support

The flow of technology to suppliers, educational institutions, maintenance and repair shops, or customers is of singular value to the affiliate in local-market development. For most manufacturing facilities, the success of their long-range programs will depend not only on the plant quality control but also on the dependability, cost effectiveness, and quality of suppliers. Local universities and training institutions receive materials and technical support in order to improve their courses that train the necessary workers or managers. In addition, most of these affiliated plants are encouraged to support local standards associations and other professional societies that stimulate improvements in the technical culture of the host environment. Some governments have just begun to identify the potential value of this type of technology, which shows a more innovative approach to security technology than the narrower policy of purchasing processes or equipment alone.

To enhance the concept of evaluating the stages and mechanisms of technology transfer, work should progress on identifying which aspects of technology are critical to different types of industry. There is no one best means for technology development and transfer. For mass-production technology, one element may be more important than another. For the recipient, the stage of design and construction may bring more critical new technology than the start-up phase. All of these determinations depend not only on the type of industrial technology being supplied, but also the overall needs of the receiving unit, its environment, own stage of development, the supportive infrastructure, and its relation to world supplies and markets.

MECHANISMS OF TRANSFER

The different mechanisms of transfer of technology indicated in the matrix given in Table 1-1 are not always applicable to each of the types of technology transferred, and their use will be more or less appropriate during different times of the life of the affiliate.

Documentation
Throughout the life of the affiliate, a number of manuals, specifications, designs, layouts, process instructions, and so forth will be supplied to the affiliate, relevant to the different stages through which it is going. These documents can be quite detailed in providing substantial "secret know-how" to the affiliate; or they may be rather general in providing only broad guidance, such as might exist in training manuals. In the latter example, the recipient is left to fill in the specifics and make a variety of decisions on his own. For the more detailed documentation, one of the most significant lessons

to be learned is that, however specific the instructions may be, they are not generally adequate enough to permit the recipient to produce precisely what is achieved in the headquarters plants. The reason for this is that drawings, specifications, designs, and so forth seldom *fully* reflect shop practice. Shop practice is what actually occurs on the plant floor in the process of manufacturing. Repeatedly, affiliates have gone to another affiliate or to the headquarters plant to ask why they cannot produce the same result following the drawings only to be told, "Oh, we don't do it just exactly that way." The modification made in one plant may not be the ones made in another; but all such shop modifications are made to achieve a given result, which is simply not producible by strictly following the specific drawings. Many of the shop foremen, engineers, or on-the-line workers know little tricks to make the operation go smoothly or to produce the desired result. Some of the supervisors or foremen or inspectors keep little black books of information as to how to adjust or modify particular processes or results in order to achieve what is wanted. Others merely keep the tricks in their minds, which is a way of assuring their own job security since no one else has that precise information.

The results of the gaps between documentation and shop practice can be quite striking. An example of many years ago is that of a company that received a cable from a Japanese firm requesting that it come to Japan to repair or fix one of its machines. The engineers arrived in Japan to see what was wrong with it. They worked on it for several days but were unable to find out what the difficulty was. The machine was certainly one of theirs with the proper trademark and serial numbers clearly evident and in the right places. After much frustrating work and cables back to the parent to find out the history of that particular machine, it was discovered that the Japanese had bought one machine, had duplicated it precisely through their own production, and had copied even the serial numbers and the trademark stamp onto the copied machine, but had not been able to make it work. Something was missing in the processes or procedures at some stage (in someone's little black book), which made the difference between merely copying and functioning.

The practice of not communicating *all* knowledge exists even among affiliates of the same company, despite the fact that there are regular reporting procedures for all affiliates on what they are doing. Reports on proposed research programs or new product developments are circulated throughout the company, but they are frequently found to be incomplete by another affiliate who might like to adopt the ideas. Not only are reports on what is being done in new products incomplete, but research proposals for new products are sufficiently incomplete as to make it unlikely that another affiliate would steal the idea and adopt it for its own. Although there are clearances for such activities at headquarters, even headquarters may not be fully apprised of the objectives of the activities by the affiliate.

The only way which these gaps can be filled is through another mechanism of transfer—personal visits—which will be discussed below.

Instruction

The mechanism of instruction tends to peak at the beginning of a project and at major product changes. It exists throughout the daily life of the affiliate, however, since there is always turnover and new workers are coming into a number of the sectors of the plant activity.

The significance of formal education is recognized by most of the companies, even for their workers, and many of them are increasingly expanding the formal education of their personnel by offering tuition, books, travel expenses, time, and a variety of encouragements to raise the basic education levels, introduce special skills, and broaden the expertise of engineers and managers. These inducements, however, frequently come a little later in the life of the affiliate. An early stage is frequently much too hectic to permit key people (workers, foremen, supervisors, engineers) to be absent during the work day. However, as the schedules settle down, the companies tend to become more interested in urging formal education and in supporting it both during and after hours. Unfortunately, the experience in several of the countries we investigated is one of a lack of eager reception on the part of workers of the programs offered or the opportunities available.

The other major form of instruction, of course, is on-the-job training. Such training, which is initiated even before the plant is in operation, uses models of the plant and early equipment, to train the supervisors and some of the key workers. Once the plant is in operation, this training becomes much more extensive and formal in the sense that it is built into the system along with preliminary training when a worker joins the plant. Such training varies substantially with the type of tasks undertaken; some tasks do not permit any on-the-job training, because errors must be kept to an absolute minimum, while others can permit errors of up to 15 to 20 percent until the worker improves to the point of a 10-percent-or-less error rate or whatever might be the average on the plant floor for that task.

Conferences and Seminars

The use of interaffiliate conferences and seminars has been found to be quite different among companies and even varies within a given company according to different departments or functions. For example, the "quality and assurance" department of a company may have extensive conferences and seminars to make sure that everybody is up to the same standards. On the other hand, research groups may not feel the need for such conferences since they make their exchanges through papers, whereas product development groups may meet annually or even more frequently. The frequency and the content of the program appears to depend very much on the individual personalities of those who head up given divisions or product lines in the company.

Again, the frequency of such meetings varies over the life of an affiliate in that affiliate managers or engineers will choose to attend such meetings depending on whether they are in a phase of new product development for new process adoption, or general expansion. At such times, they are much more interested in what is happening elsewhere so they might adopt it. Otherwise, they may find that attendance is rather expensive and can be by-passed. Also, of course, general economic conditions or sales volumes may dictate whether or not the affiliate feels it can afford to send members to particular conferences.

In some instances, the conferences and seminars are given regularly, with the agenda sent to the affiliates to let them decide whether or not to come. In other cases, they are mandatory conferences set up for a given purpose and all must attend.

Visits and Exchanges

The movement of specific individuals among affiliates or between them and headquarters has at least three different objectives. One is for the headquarters officials to review what is going on in the affiliates and to carry information to them about what others are doing. A second objective is for the affiliates to search out what others are doing that they might adopt, which cannot be gained from either conferences and seminars or reporting procedures. The third is preparation of an official in an affiliate for a new job, or preparation of a new man for an existing task. All of these will be found in the cases reviewed in the succeeding chapters, but each company emphasizes the significance of such visits rather differently. In one company, such visits are the key element of information exchange, despite rather extensive programs of documentation and conferences and seminars.

Exchanges of personnel among affiliates for the purpose of job training occur for periods of up to three years, with the individual official possibly serving in more than one affiliate. Not only is this useful in terms of the interest of the sending affiliate, but the receiving affiliate also gets some information exchange, and both eventually profit from the ties that are developed through this particular individual and that facilitate future exchanges of information.

Casual visits can produce some very striking results; for example, surfacing information that simply was not available otherwise and making significant contributions to the productivity or changes in product line of the given affiliate.

Both visits and exchanges are unlikely to produce the same results between a licensor and licensee that they do under the 100-percent-relationship.

Equipment

The sending of new equipment, or of any specific equipment system, to an affiliate requires a transfer of know-how with it. Not only must the recipient become familiar with the design and the objectives of the machine, but if the equipment is to be bought locally in whole or in part, they must understand contracting procedure, installation requirements, the precise function and use of the machine, its maintenance and repair, and potentially its modification in case of inadequacies or shifts to new product models within the plant. Little of this can be transferred to the affiliate by documentation, despite the numerous documents that would be passed from the headquarters company to the affiliate. Oral communication is required along with the transfer of the equipment, in the form of a specialist who can oversee installation and explain the functions, maintenance and repair, and so forth.

This type of assistance is also sporadic, since it is related to the adoption of new machinery and usually goes along with a change in the product line. However, there are also breakdowns and problems in maintenance and repair, as well as modifications, that require communication of techniques.

Trouble Shooting

One of the major advantages of association with an international company is its ability to send individuals or teams of experts to cure a particular problem—one that may have arisen in a given plant only or that has occurred in one or more plants and requires handling in others as well. The expertise of such individuals or teams lies in a quick identification of the problem and an application of solution-techniques that comes from long experience and from rigorous and logical thinking. Without a careful identification of the problem and the precise solutions necessary, what is frequently employed is a trial-and-error method, which frequently introduces new problems while curing the existing one. In several instances in the cases studied, we found the local affiliate engineers solving one problem only to generate others, while not recognizing the cause of either.

The absence of training in problem analysis and solution is a critical gap in the competence of many workers and supervisors in developing countries. If a given company received only a license to produce a particular product—largely out of documentation—this particular type of technology transfer would be missing. While license contracts can cover it, there is always a reluctance on the part of a licensor to commit a highly valuable team to a given problem of a licensee, when that problem might be more readily solved by less expert individuals. The licensee may merely have misidentified the problem, thereby causing large expenditures of manpower and time for nothing.

Related to the transfer of technology through equipment and the trouble shooting aspect is the reduction of "down-time" for machinery through its proper installation, use, repair, and application within the production system. Equipment, with layouts designed by American manufacturers, is frequently desired around the world above others because of the fact that the machinery stays in operation longer—that is, the percentage of time loss for breakdowns (down-time) is lower than for other equipment or systems. Although this equipment and some of the layouts can be purchased independently, again, the "little-black-book" syndrome remains critical; getting the maximum performance out of the machinery or the layout in the production process requires the "secrets" discovered by the parent company.

In sum, no one of the above mechanisms is adequate to transfer all of the technology necessary by an affiliate, nor is it possible to program the precise amounts of each needed at given times prior to the fact of operation. The different mechanisms will be used in quite varying degrees by different affiliates and over different periods of time. Only with the closest of business associations (a wholly-owned relationship) can an affiliate expect to tap fully the technology of the parent company.

SIMILARITIES IN TRANSFER PROCESSES

Despite the fact that there is no one best "technology package" for firms in any industrial sector—nor best means of transfer—there are some similarities among the patterns found in the companies studied here. One of the first similarities encountered was that of an initial reluctance to have their stories on technology transfers told; yet after going through the experience, they have concluded that any fears of the results were unfounded. Many officials now think that more such investigations would be helpful, by providing detailed information within a real setting. Such data are, of course, more useful for policy prescriptions than "aggregate data" from several dissimilar companies; for it is increasingly clear that in the arena of international investment, disaggregation is necessary if appropriate policies are to be developed.

A second similarity among these cases is that full technology transfers would have been highly unlikely without at least majority ownership, and probably full ownership, simply because of the intricate relationships among affiliates, markets served, and nature of the technology. In none of them did any official consider that it would be possible to obtain similar (and certainly not identical) technology from outside of the company.

In each of the cases, technical transfers were continuous—that is, occurring over a wide range of activities and throughout the life of the association. There has been no clear delimitation on what technology could or could not be transferred; rather, all the data and know-how necessary to do the job was transferred, not always without discussion, however. Since there is no precise delineation or even control over the technology transferred, there is no way in these cases to cost-out the assistance provided. Each company has a different method of charging for technology supplied, but none has developed precise cost-price methods that reflect the underlying value of the wide range of techniques, processes, specifications, and know-how provided the affiliates. Cost allocations are necessarily arbitrary; therefore, only "out-of-pocket" charges are readily allocable. The value to the receiving country is also impossible to calculate without precise data bases—that is, any value set is either arbitrary or merely the result of negotiation among willing parties.

In all cases included here, the determination of "appropriate" technology fell out of the products chosen to be manufactured, the processes already developed by the parent or another member of the enterprise, the support capabilities of the parent or affiliate acting as "mother company," and the learning capabilities of the affiliate receiving the technology. These determinations were modified in each instance by scale factors, which required some processes to be de-skilled or de-automated into manual operations. But no "principles" of determination of appropriate technology were discovered among these cases.

Finally, it is clear from these cases that countries differ in their ability to receive and absorb technologies at different levels of sophistication. In several of the cases, the technologies had to be deskilled so as to reach the educational levels of the personnel available. In others, extra educational efforts were required to train the workers up to levels where they could absorb the technologies required. These educational programs included the sponsoring of programs outside of the company, as well as inside, so as to develop supervisory personnel and those who could eventually rise to managerial positions.

A signal dissimilarity among these cases is seen in the organizational structures used to support or channel the technology transfers. Line and staff relationships are mixed differently, and the significance of personal visits varies among them. There is, therefore, no single pattern that will necessarily succeed though some procedures are better than others.

CONCLUSION

The difficulty and confusion surrounding technology analysis and planning stems from an incomplete view of the process of technology development and transfer. Manufacturing technology is supplied for different stages of plant development and environmental support of the new plant. Each technology varies according to the stage, and the unique characteristics of its industrial classification. Guidelines are emerging that aid in evaluating the mechanisms of transfer, the technology for different stages, and the relative importance of these stages to varying manufacturing activities such as mass production, unit production, or continuous process.

Effective planning for both the supplier and receiver must begin with identifying the full scope of technology that supports a manufacturing process and then by identifying which stages are more important to which types of activities. Early detailed planning and joint participation in all stages will prevent confusion and improve the impact of the transferrable technology. It will also change the focus of technology debate from that of analyzing independent component parts in isolation to that of a positive approach of planning for all possible technology benefits and combinations of components within a complete system of technology development and transfer.

 *Part I*

Ford Motor Company

Introduction to Part I

Ford has been in international business almost since its inception in 1903; by 1920 it had established plants in England, the European Continent, Scandinavia, and South America. It manufactures and assembles in 29 nations on six continents and sells vehicles in over 200 countries and territories. Nearly half of its total employees are in overseas operations, which account for nearly one-third of its car and truck production. Conducting these far-flung activities requires continuous interchange of people and technology between the parent company and affiliates and among affiliates which involves vast correspondence and numerous personal visits.

In order to assist affiliates in developing countries to accept, understand, and utilize the flow of information, Ford conducts extensive technical-assistance programs and training efforts, often by helping to develop a local supplier industry so that it can meet the exacting quality standards required. As a result, thousands of mechanics and skilled tradesmen have been trained not only for Ford's own operations but also for customer service and indirectly for employment in supporting and competing companies.

To tell the story of the technology transfers would require a minute account of daily activities of hundreds of officials of the company and their contacts not only with each other but with down-the-line supervisors and employees. Most of these interchanges are not recorded nor costed by the parent or affiliates; some of them are, however, and fairly precise accounts are kept during the programming of a new facility. A roughly accurate picture can, therefore, be drawn by tracing the organizational relations, the flow of requests and assistance, and the impacts of technology transfers within affiliates and to their suppliers or customers. This is done here for two of Ford's affiliates: South Africa and Taiwan. To understand the scope and nature of these transfers more fully requires, first, an appreciation of the headquarters and regional organizations from which they come.

Chapter 2

Management of Technology Transfers

The flow of technology to any affiliate comes from a variety of sources—headquarters in Dearborn, a regional staff, and other affiliates. But the principal resources available to an affiliate come from or through headquarters and the regional management centers.

WORLD HEADQUARTERS

Ford's international automotive operations are headed by an executive vice president, under which there are three regional companies, located in Mexico, Britain, and Australia (see Figure 2-1). Serving this vice president is a staff official in Overseas Business Planning, and backstopping all of his overseas operations is the entire staff of the headquarters company in Dearborn. For purposes of describing the process of technology transfers we can omit discussion of the financial, public relations, and other nonmanufacturing staff functions, which report to the chairman of the board. Of direct significance for these transfers are those functions that relate to product planning and design, to plant layout, construction, and launching, manufacturing assistance, supply, training, and dealer service. Most of these functions fall under an executive vice president–operations staff, and a vice president for product design, both of whom report to the company president. As of mid-1974, the operations staff included units responsible for Technical Affairs (manufacturing staff, supply staff, and scientific research), Environmental and Safety Engineering, and Product Planning and Research.

27

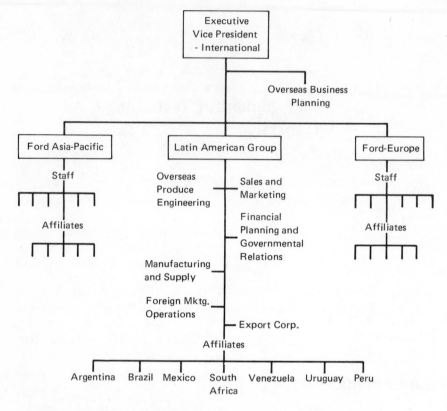

Figure 2-1. Ford Regional Organization.

Industrial Engineering

No operation is set up overseas without full participation of the headquarters staff in the design of the plant capacity, layout, facilities, and equipment. The Bordeaux (France) transmission plant was considered one of the most successful plant launchings, partly because it was completely planned in Dearborn, by drawing from worldwide company experience and by relying on personnel from the European affiliates who could perform special tasks. The Ipananga plant in Brazil was copied directly from the Lima, Ohio, engine plant, which made it a highly successful venture.

Many plants in less-advanced countries draw from layouts and methods of a few decades back, with modifications for the adoption of more modern techniques to reduce costs. The Office of Manufacturing Engineering and Technical Systems designed the layout for the Brazilian foundry, from start to finish, largely by adapting U.S. methods but incorporating a unit for a smaller volume of molten metal (grey iron) production, using the induction melting process, rather than cupola. This process provided flexibility as well as a high degree of specificity in the composition of metals produced. It had not been used before by Ford, but its success became the basis for identical facilities in Mexico and Argentina, and the technology was later adopted within the U.S. plants.

Given the fact that much technology transferred abroad comes directly out of experience somewhere else in the company—as with the induction process—it is usually difficult to determine the aliquote cost of such transfers to any one affiliate. However, in the case of the induction process transferred to Brazil, since it was begun from scratch by Ford-Dearborn, the project was a line-item in the budget and precise records were kept on man-hours spent in the gestation and transfer. Still, calculation of total costs would have had to include a portion of corporate overhead (40 to 50 percent of the time of the manufacturing staff alone), and the expertise drawn on here would not be costable.

By estimates of the staff, the "launching cost" of a well-engineered new plant is between 10 and 15 percent of the cost of the capital equipment used. Thus, a $100 million investment in equipment for a new plant will require between $10 and $15 million of pure technology transfers. If *all* technology for such a plant had to be developed from scratch, it would cost about $70 million. (By comparison, the "from scratch" cost of the technology to build a $100 million plant for development of oil shale would exceed $2 billion—but, of course, one cannot cost-out the base of the technology pyramid upon which new developments are built.)

The staff has attempted to put a value on moulding specifications for foundries and has calculated it at 10¢/ton/year in modifying existing green-sand moulding techniques. (Of course, the changes come in lumps with technical breakthroughs and are built on discoveries by all moulders around the world.) The cost of developing and transferring continuing technology in moulding is estimated on the order of 2–10 percent of the value of production per year.

The value and complexity of technology transfers have been found to vary directly with scale of production—large volumes and economies of scale permit technologies for automated production, with less cross-loading of parts on the assembly lines. The trade-offs among men-capabilities, machine-capabilities, and costs determine the types and extent of transfers of technology. An effort is made to study the costs of engineers/hour for the technologies transferred. Over the life of any one plant, these technologies shift, for example, South Africa's Ford plant now looks like those in Mexico and Brazil did some years ago, and it will change as production increases and product development shifts. These technical changes are reviewed by the manufacturing staff before top management is requested to agree; any changes are costed-out before modifications are made in feed, speed, transfer lines, stamping, paint or whatever.

Product Development

The product development staff preprograms any new vehicle. The conceptualization of the vehicles to be produced in a foreign affiliate is the result of cooperative efforts guided by the Dearborn staff, taking into account markets, production capabilities, workers and skills available, materials, and so forth. The engineering facility with the product development staff has responsibility for engineering any new products that are to be manufactured overseas; so the new Bobcat, to be made in Spain, was designed and developed in Dearborn along with the hardware needed for production. The extent of these developments dictated the volume and nature of the technology transferred. To determine the design and configuration of the Bobcat, the staff took a number of competitive models apart and determined what a "best composite" would be like if it were to serve multiple markets around the world. New parts and components had to be designed, which require technology transfers to each affiliate and its suppliers. Six U.S. personnel were transferred to Europe for the project.

Ford-Europe did the product development after Dearborn approval of preprogramming and the production engineering. The Bobcat cost $5.5 million in product development alone, including hourly wages and salaries of personnel, materials, support services by the company, and overhead charges. The design project required eighteen months, of which twelve were in research on alternate models. Modifications after these prototype tests mean that the final model will be different, engineering-wise, from the original Dearborn model. The long time required to complete the model stemmed from Ford's concern to achieve leadership in the chosen markets with the help of low-cost manufacturing.

This task could have been accomplished only by an existing auto company—not by a newly formed one relying on outside technical assistance, for it could not find the same type of personnel to develop a similar product. To achieve such a task, an organization is required that can simultaneously develop a vehicle and specifications for components. South Korea is attempting to design its own car and will do so by copying someone else's; there is little difficulty in doing this, and the car will be satisfactory for a closed market.

What is difficult is to design a *new* car which requires engineering of new parts and components. The process also requires close liaison with manufacturing engineering and industrial engineering to minimize the costs of production and assembly; both will offer suggestions on component design to minimize costs and processing or machining. Conversely, plant layout is affected by car design; for example, following European practice, Ford has begun to put the engine in from underneath along with the suspension in a unitized system; this cuts two inches off the length of the car and saves considerable manufacturing costs (thereby offsetting the $4.5 million cost of the research project to sketch and cost-out the changes).

The car design staff gets its responsibilities assigned after a planning staff determines how to fit the new car into the total models of the company and where it should be built. The Dearborn engineering staff completes its work with the preprogramming and acceptance of the program; Ford–Europe's engineering staff has completed its task with the production of "job-one" in Europe. The Dearborn staff is composed of 150 expert American engineers; some have had foreign assignments and are therefore familiar with production problems abroad. The design staff operates in conjunction with other staffs, which are responsible for research on alternate engines, for the business planning aspects of new products and new facilities, and for longer-range planning of car and truck components.

A product planning and research staff, in the operational staff group, reviews any proposals for new products or product changes by any affiliate. However, the affiliate is not bound by the results of this review. The group will represent its view in a technical assessment and continue negotiations with the affiliate to achieve a resolution of differences, if possible; if not, the technical assessment is passed up the line for decision. For example, in the development of a recent new model, dialogues have been undertaken with Ford–Australia so as to include their demand and production needs; an Australian engineer joined the Dearborn staff for a time to bring his knowledge of sealing problems to keep out hot, blowing sand and cold rain. In other instances, European engineers have contributed their experience with weight reduction and have sometimes brought a different engineering orientation to the discussions.

To assist affiliates around the world, there is also an Overseas Product Engineering Office with 300 personnel. Product engineering services are provided, especially to those not having a sufficient staff of their own (such as in Brazil).

Manufacturing Staff

Once in operation, the foreign affiliate has the entire technical capabilities of the Ford company at its disposal in resolving specific problems of production or assembly. Product or component units among the affiliates initiate requests to the manufacturing staff for assistance; for example, in painting, metal finishing, stamping, or whatever. In Australia, the suspension designed for the Falcon did not stand up because of the off-road use (similar to the 1920s in the United States or to a pick-up truck in the West today); the suspension was high enough off the road level but not rugged enough, so a request was made to the manufacturing staff for modification.

Similarly, Brazil has had problems with corrosion resistance. Ford had solved these problems in welding during the 1960s, and though it had informed all affiliates, many did not adopt the technique because they did not think they had such a problem. Once Brazil developed the problem, it turned to Dearborn for assistance and was informed of the zinc-rich welding process; it is now obtaining zinc-rolled-coat-metal for this process through Diamond-Shamrock's affiliate in Brazil, though production is just beginning there.

These requests come directly into the director of manufacturing engineering and quality control in Dearborn from the manufacturing manager in the affiliate. This direct line is encouraged and is used because the manufacturing staff has functional responsibility for all plant technology and processes; all top operating officials abroad have been to Dearborn for visits, so personal relationships have been developed that make direct contact easy and informal.

In addition, the more than forty experts on this staff (all with over ten years of experience) have themselves been overseas for short visits, and ten have had resident assignments abroad. These officials have only an advisory position in that the affiliates have the final decision, but the staff will press hard if it considers its views are critical.

The manufacturing staff has departments related to machining processing, stamping, assembly, forging, and foundry and steel-making; this staff is able to draw on complementary staffs responsible for plant engineering, for development of plastic products and components, and for development of all other manufactured products.

The manufacturing staff is also a training ground for foreign managers; initially, manufacturing managers of affiliates often are sent out from U.S. operational positions, but some are brought up through the ranks of other affiliates and are given guidance from

U.S. officials during the launch. Those coming to the United States will be trained for six weeks to two years with the staff departments, as technicians, working on specific processes and machinery to obtain complete familiarity with the system and personnel. Over the years, between five to ten officials in each foreign affiliate would have had this type of training in Dearborn, with the manufacturing manager having it at an early stage of the launch and prior to his promotion to that post. The cost of such training is high: $2 million for thirty trainees during a single year, plus the time-cost of the Dearborn staff in teaching. In one group of thirty in 1974, there were twelve from the Brazilian plant alone; prior to this group, there were twenty Brazilians up from the engine plant for training, which facilitated the direct translation of the Lima, Ohio, engine plant into a new Brazilian facility for 2.3 litre engines using high technology.

Although the foreign affiliates often seek assistance directly from particular facilities and personnel in the Dearborn plant operations, these divisions cannot oblige without an approval from the manufacturing staff, for funds and man-hours are required. The specific needs of the requesting affiliate will be checked out before an "OK" is forthcoming.

To illustrate the extent of the involvement of this staff, in the launch of Ford-Lio Ho (Taiwan), the staff did the preprogramming in Dearborn—that is, preliminary estimate of costs (facility costs and unit costs) based on formulae developed from experience, man-hours required, equipment needed, facilities, service support, and so forth. This required some nine months of work for four men, or three man-years. The program was sent to Ford–Asia–Pacific (FASPAC), the regional management center in Australia, which added detail and returned the package to Dearborn for checking; three professionals went along to explain the detail and work out any kinks found in the equipment, layout, cost estimates, manning requirements, and so forth. In addition, Dearborn has a man at FASPAC each year checking on problems in the area; he provides liaison as needed and helps to determine the kinds of skills needed by an affiliate in that region. The experts might be drawn from anywhere, but probably from Britain; and on three occasions, two Ford–Britain individuals came through Dearborn on the way to Taiwan for discussions and clarifications. .

The cost of this assistance is "donated" to all affiliates, with affiliates paying only out-of-pocket expenses of visitors. Nearly 50 percent of the time of the staff is devoted to foreign affiliates. In addition, some 90 percent of U.S. studies and work is directly usable abroad in some affiliate, so that foreign engineering in developing countries consists of adaptations and "re-do's" of earlier operations in the United States; for example, circa 1940s for Ford-Lio Ho in Taiwan. The Brazilian engine plant was originally set up on a 1940 model Ford plant, producing 18/hour instead of 125/hour; however, it used technology for 125/hour and cut down the scale by segmenting the line and shifting workers from one phase to the next moving along with the engines; eight breaks were introduced and all workers were trained for each step. The result was a set of trained mechanics rather than skilled production workers. Production can be increased merely by reducing the breaks in the line.

Training of manufacturing technicians and managers occurs at worldwide workshops, such as the one in Mexico in October 1974 (the first such session for Ford). It was attended by industrial and process engineers from Mexico, Brazil, Argentina, Australia, South Africa, Europe, and the United States. In addition, industrial engineering meetings have been held on specific problems; for example, "fastener technology" including nuts, bolts, and screws of which 37 billion are used per year. These workshops are taken to each affiliate.

Results of this extended assistance are seen in recommendations to change vendor/suppliers of affiliates to improve quality of tools and raise productivity; in promotion of local suppliers of items such as torque wrenches; and in a Swedish supplier's proposal to the International Standards Organization that it adopt Ford standards worldwide.

Supply Staff

Within the supply staff, an overseas procurement planning unit provides liaison for all foreign affiliates. At least one procurement officer in each affiliate has had a tour at Dearborn or one of the United States facilities (for up to a year) and has spent considerable time in several purchasing units within the Dearborn operations. Several of the Dearborn officials have, in turn, visited foreign operations to assist in the selection of vendors for those affiliates.

The supply staff has responsibility for vendor selection, scheduling of movement of materials to affiliates in the United States, procurement of raw materials, purchasing of facilities and machinery, and transportation of supplies. This expertise is available to these responsible for liaison with the foreign affiliates. Officials of this staff participate in the determination of what items will be procured locally by a foreign affiliate. For example, this staff studied the requirement imposed by the Peruvian government on local parts procurement and decided, on that basis, not to bid for a role in that country's auto industry.

This staff is involved in the decisions on any new foreign investment from the standpoints of available vendors, supply capacities, scheduling of production, and delivery of parts, components, and materials. After the affiliate is in operation, there is continuing need for consultation with the supply staff, especially on the timing of changeovers to new models; in addition, the staff is available on call from any affiliate.

Three liaison officers are available full-time and can answer many of the questions directed to Dearborn, but in addition some 25 percent of the time of the director is spent overseas with affiliates. Much time is also devoted to consulting with U.S. vendors to obtain help for foreign suppliers of affiliates, with the charges paid by the affiliate. About one-third of the time of the staff is spent as an intermediary between potential U.S. licensors (suppliers to Dearborn) and foreign vendors (potential licensees). In this broker function, Ford will try to. get a U.S. vendor interested in licensing abroad but will not try to select the specific licensees. This activity is quite significant in some countries and is particularly important in specific items, such as brakes.

Training

As in the supply staff, within the personnel and organization staff are two liaison officers who have responsibility for continuing contacts with foreign affiliates—one in the Education Planning Section is responsible for contacts on training programs and another in Management Personnel Planning for requests for specific individuals to be sent out for training purposes and personnel guidance. (In addition, Ford has within a General Services Division a Management and Technical Training Department that provides, at cost, materials for specific courses on request from the affiliates.)

Out of its early experience in Canada, the company has developed an extensive Planning Guide for Training. It is a detailed program to show affiliates how to draw on technical expertise in training workers and supervisors. Its table of contents describes over eighty pages relating to planning and implementing training, problem areas, writing job tasks, learning modules, selecting training resources, evaluating training programs, and specific programs for different levels of responsibility. This guide is credited with helping to smooth the launches of new operations overseas. For example, the Bordeaux plant was considered a very successful launch partly because the

German plant manager helping to launch it was in Dearborn for two years prior for discussions on many aspects including training, and people were sent from the Transmission Division to Bordeaux for forward planning, using video tapes of the Lamm transfer machines (a complex 40-station machine for tooling transmissions), on metal stamping, and on other processes.

Similarly, many Brazilians were trained in the engine plant in Dearborn; and three individuals were sent from general services as "contract trainers" for a year. Although these last are paid by the affiliate, the time of all others is donated and included in the cost to the parent company of a launch. The foreign affiliates, therefore, get materials free that have been developed for U.S. and other foreign affiliates. However, if they wish one of the "learning programs" developed by General Services, they must pay a fixed fee, which is still below costs; for example, the English Logic Static Control program can be bought for $4,000 (though it cost over $18,000). Many of these programs employ video tapes, and such cassettes, which are increasingly being used for the exchange of information among affiliates, relate to processes, repair and maintenance, testing, and so forth. There are literally tens of thousands of such cassettes available to the affiliates from various sectors within the company.

Training is also accomplished through a variety of worldwide conferences. For example, in 1972 a conference was held in Houston for salary-personnel managers and training managers; fifteen to twenty overseas officials attended the five-day sessions on the use of video tapes, interpersonnal skills, and the experiences in Europe and FASPAC. International workshops are held on specific developments, such as electrical systems, at the initiative of Ford-Dearborn. The affiliates select the technicians to attend and pay their travel costs. In addition, Ford-Dearborn sends task forces to affiliates to meet specific problems, such as sealing, electrical warranties, and so forth, as requested by affiliates or at the initiative of Dearborn, which watches for problems.

REGIONAL MANAGEMENT CENTERS

The three regional management centers also have technical staffs that support the operations of the affiliates located in their area and even some of those outside, as a result of product similarities. Thus, although South Africa comes under Latin America for managerial and reporting purposes, it is served largely out of Ford-Europe because it produces British-model cars and trucks. The regions are organized as shown in Figure 2-2, each with adequate

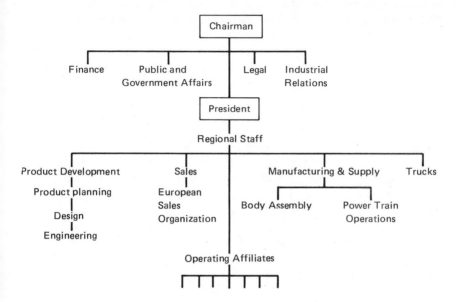

Figure 2-2. Organization of Ford-Europe.

staff to support continuing operations and with ties to World Headquarters if they cannot meet a given problem.

Ford-Europe

Ford of Europe, Inc., is a legal entity but with no operations of its own. It coordinates all operations in Europe, those in Africa (save South Africa), and those in the Mid-East from Pakistan toward the West. Although all functional personnel in the operating affiliates in Europe report *directly* to their own company top management, many wear two hats and are responsible for coordination of affairs in their own area of responsibility outside of their own company; these personnel report *functionally* to Ford of Europe. Ford-Europe is considered a profit center for company purposes, but since it has no income, its personnel are on the payroll of one affiliate or another throughout the region.

As can be seen in Figure 2-2 the organization of Ford-Europe shows the control channels and the extensive support resources available to each affiliate. Each of these staff units has direct line coordination over similar functional units in each country affiliate. The primary task at present of the Product Development unit is to develop a single European line, encompassing the Granada, Escort, Capri, Cortina, and commercial vehicles. Product selection, which is a joint determination of local companies, regional headquarters, and Dearborn, takes into account national preferences and tastes. Technical assistance is provided primarily through the Product Development and the Manufacturing and Supply units. Such assistance is given even to U.S. units handling the Capri; they receive instructions on repairs and maintenance and help on instrumentation and tool supply.

Ford-Europe, in turn, is assisted at times by Ford-Dearborn personnel, on such problems as upholstery, paint, and materials handling (of CKD units). Given the importance of rapid communications (piles of Telexes are received and transmitted by each regional headquarters on manufacturing problems), all units are tied into a worldwide communications network, and Dearborn provides assistance in communications technology and computers. Personnel are also drawn from retired officials of Ford-Dearborn, Ford-Europe, and Ford-Australia to staff particular projects or study groups examining a potential new project; these persons are especially useful as cost estimators—not only because of their experience but also because they have no personal stake in the results.

Technical personnel are also sent out from independent suppliers of U.S. affiliates of Ford and from Ford affiliates that are suppliers to local and foreign affiliates. For example, a machinery supplier will be sent out to an affiliate in connection with a new layout or a production problem. Ford's U.K. affiliate producing diesels will send technicians throughout the world to aid those assembling and servicing diesels. (Ford-UK concentrates on research on diesel engines, while emissions controls are largely the responsibility of the U.S. research unit.) The U.K. affiliate also produces the smallest engines and the lightest transmissions of Ford's worldwide operations and thus extends technical assistance to all relying on these components. For example, the New Zealand operation set up a manual transmissions plant two years ago. It first assembled U.K. components; it bought castings and forgings and it ran into substantial quality control problems that required a visit from a British metallurgical expert. At the third stage, the U.K. company sent a team to set up a plant, lay it out, and start it up; assistance was also provided by process engineers, and a "foreign service" officer was loaned to help get the operation started successfully.

There were some 495 such "foreign service" officers detailed from one affiliate to another in 1974, though this is expected to drop to around 300 in 1975. Their salaries are paid by their own company, but they get "field costs," including moving costs, from the recipient company, which also reimburses the lending company for full salary costs. Ford-Europe itself will be reimbursed for technical assistance to South Africa or the Far East for "out-of-pocket" expenses only—that is, there is no time apportionment of basic salary of its officials—unless the visit is over six months long. Such assistance would normally be provided only at the request of the executive vice president–international, or of the manufacturing staff at Dearborn, or the president of the Latin American Group. European affiliates, which are tied to South Africa through sales of components will, of course, supply technical assistance directly and receive their returns through their exports, which include a charge for know-how transfers. Ford-Dearborn assesses each affiliate an amount based on its percentage in total travel costs of Dearborn personnel.

The transfer of technology is enhanced through direct person-to-person contacts derived from the many functional meetings sponsored by Dearborn or the regional headquarters. Out of these come not only technical exchanges but also better communication among officials who have to correspond on a variety of problems.

Technical assistance is also extended to suppliers; when inspection under a "supplier quality assurance" program demonstrates a need, technicians and engineers are sent out to assist in rectifying problems in production and quality control. The suppliers are given advice on manufacturing processes, management, cost-control, and so forth to keep them on their toes competitively and to keep Ford's costs down. Several British suppliers have admitted that Ford, as a customer, had kept their companies alive competitively. In addition, suppliers from the United States are induced to expand abroad; some fifty U.S. companies were encouraged to set up in Australia, with Ford assistance.

The technical assistance available to any affiliate is reflected in that given to the transmission plant at Bordeau, France. Ford-Dearborn determined a need for auto transmissions that were small, light, and different; Ford-Europe fed in substantial demand data and requested manufacture in Europe, based on cost data. The transmissions were designed in the United States. Site selection was based on comparisons between France, Britain, and Germany. The decision went to Bordeaux based on a number of considerations, including personal persuasion by de Gaulle during a visit of Henry Ford, II, the existence of a port (facilitating shipment of two-thirds of production to the United States), high quality and available labor, good infrastructure, and proximity to a technical university.

The plant design, which was done at Ford-Cologne, drew from the Ford Plant Engineering Manual, which details power needs, load factors, and so forth. The project manager for the plant was detailed from Ford–Germany. The layout, processing, and tooling were duplicated from Ford–US plus all the improvements that had been developed but not yet implemented there. Bordeaux, therefore, obtained the absolute latest in developments, which U.S. engineers would like to have had but could not yet afford to adopt. Design and production engineers were seconded from Britain and the United States; they made 894 design changes in the original U.S. specifications before final production, based on testing of prototypes driven by company officials on daily business. Other personnel were assigned from the Transmission and Chassis Division in Dearborn. As a result of such cooperation, the launch of this plant was considered the smoothest of any in Ford history.

Once in operation, technical assistance from Ford–Europe continues to flow in a variety of channels and directions. For example, a machine was recently developed to clean the castings for transmissions housing, by keeping lubrication free of filings and scrappings; it will be used in many locations. A solution is presently being sought for piston-scuffing in engines in Britain, which has developed recently after twenty years in production. Australian production of Cortina and Escort (85 and 70 percent local content, respectively) is assisted by the transfer from the United Kingdom of every change in specifications and drawings through a resident engineer in the United Kingdom. South Africa also has a resident engineer in the U.K. plants making these autos who notes changes and adopts those desired by Australia. Ford–Australia is expanding the manufacture and use of plastics, based on U.S. technology. Two Australians were sent to the United States and slowly built up production in the plant by introducing new items into the design of the autos. In addition, emissions standards and the technology for meeting them are transferred, as is technology for the economical handling of materials and for reducing the weight of materials, thereby increasing strength, improving styling, and cutting costs.

All such assistance is available to any affiliate over the world, but especially to those like South Africa who use European models, whether or not they are within the responsibility of the European headquarters and despite the fact that it bears some of the costs of these transfers with recompense. For example, engine standards and specifications and test procedures are disseminated to all affiliates, everywhere around the world.

Ford-Asia-Pacific

Similar to Ford–Europe is the regional organization for Asia-Pacific, which was responsible for the gestation of Ford Lio Ho in

Taiwan. Since Ford-Lio Ho is a new project, it was carried for some time by Ford-Asia-Pacific, Inc. (FASPAC). The former president of FASPAC, his staff director for Business Development, and the present president of Ford-Lio Ho (a Chinese-American with twelve years of Ford overseas experience) were the central team for negotiating the joint venture with the former Toyota licensee; and the last became general manager of the new affiliate. All sections of the FASPAC staff were involved in the formation and gestation of the project according to their responsibilities.

As can be seen in Figure 2-3, the FASPAC staff responsibilities extend into each of the nine affiliates in the region: Japan (2 companies), Australia (2 companies), New Zealand, Philippines, Singapore, Taiwan, and Thailand. The two Australian operations and one Japanese company report directly to the president of FASPAC, however, and the other six (including a transmission company in Japan) report to a regional director, who also has a staff director of Business Development who is in charge of the production-complementation program among the five minor companies of the area.

Many of the men involved in the Taiwan project were detailed from Ford-Australia, and several are still assigned there. The plant layout, which was provided by FASPAC, expanded lines and space to accommodate the future increase in number of models and volume of production from the previous 26 per day to 40 per day. Plant engineering was redone and called for putting tools in their proper places, installing machines and electrical lines and equipment (most of which had to be redone completely), and interfacing the tooling. A maintenance engineer stayed for 18 months and assisted in everything from dormitories for workers, to office expansion, to

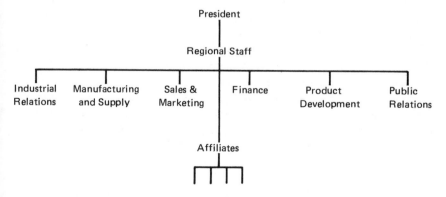

Figure 2-3. Organization of Ford-Asia-Pacific.

the production line. Over the first two years of the program, 35 individuals were seconded from the United States, Philippines, South Africa, Canada, Australia, New Zealand, and Britain, with the last ten remaining through the third quarter of 1974. Ford Lio Ho pays the bill if the visit is longer than one month. In addition, a number of visits have been made by FASPAC staff over the two years, with several officers visiting two or three times each year.

To assist the president of Ford-Lio Ho, a project manager was detailed from FASPAC (out of Ford-Australia) to oversee all manufacturing. The background of the project manager is illustrative of the expertise that is available to an affiliate of a worldwide operation. He joined Ford-Australia in 1940; was plant manager at Perth for six years; plant manager at Sydney, six years; general manufacturing manager in the Philippines for two years; and then was in Korea, Thailand, and Indonesia on special assignments to get up production volume in assembly operations without sophisticated machinery. In his wake, arrived a supply manager (detailed as a "Foreign Specialist" for an assignment between six and twelve months) to contract with each local supplier and to work with a purchasing engineer who confirmed acceptability of local supplies and provided technical assistance when needed by the supplier. Later, a number of process engineers for the assembly plant were also detailed from FASPAC staff. In addition, three British and three Australian assistants were detailed to work with the foundry and engine plants.

The project manager was the first on the scene (and was still there two years later as a "Foreign Service Employee"—meaning on "detail for over a year"). His first task was to dispel the uncertainty in the minds of the Chinese as to their future and the objectives of Ford; this took six weeks. He was soon followed by a supply officer from FASPAC (detailed as a "Foreign Traveller"—meaning on "assignment for less than six months") who determined local content requirements and went to Britain to arrange the CKD packs and deletions. (Some of the members of this team had worked together in Korea on Ford's Dealer Assembler Program for that market.) Figure 2-4 shows the program engineering phasing during 1973-74 and specifies each foreign assignment to Ford-Lio Ho.

The foreign service staff worked on the opening of the plant against tight deadlines and in quite difficult conditions: they traversed a poor road 20 miles twice daily; they often worked 12 hours per day, including Saturday; and without Western food facilities, they carried a sandwich with them each day. Over several months, this became a bit wearing.

The areas in which assistance has been received are those of plant engineering, manufacturing processes, supply, tooling, welding, materials handling, quality control, foundry, and packaging. For example, a FASPAC engineer helped with the tooling and the bid assessment from a supplier of seat frames; once the bid was accepted, he helped with the process of assembly to reduce movements and improve efficiency in handling, inventory stockage, and tool use; he was backed up by the quality-control engineer at all steps.

Assistance was received from Dearborn on request from the project manager in the form of fifteen volumes of 2-½ inch thick loose-leaf binders that set forth Ford standards on engineering, tooling, parts, paints, and so forth, plus construction of buildings, layouts, access roads, support services, and so on. When these standards are followed, there are better opportunities for assistance later if a problem arises, for others with the same standards will probably have run into similar problems. It also helps to cost-out any project more readily and to explain costs to management. Further, decisions will be more readily integrated, for the existing standards are already interfaced. And, finally, decision time is reduced substantially by following tried-and-proven methods. Not much more assistance was expected from Dearborn since it is not attuned to such small-scale operations as required in Taiwan.

The line of requests for technical assistance goes from the president of Ford-Lio Ho to the regional director in FASPAC, with copies to the president of FASPAC if necessary and to World Headquarters at Dearborn for information only. It is up to FASPAC to direct him to the proper person or solution. The lines controlling technology are fairly tight; Ford-Lio Ho can make minor changes, but concurrence of both FASPAC and the British affiliate from which the autos come is required when the change is at all significant, for there may be experience at other affiliates that would be useful, or performance may be altered undesirably.

Full functional responsibility for the manufacturing layout and timetable rests with FASPAC's staff director of manufacturing and supply. (Prior to the formation of FASPAC, the staff director of manufacturing was manufacturing manager at Ford-Australia and had a staff of six to help sister companies in the region.) Consequently, he is in Taipei several times a year to check on progress and problems. He assessed the foundry and engine plant in mid-1974 as in good shape, with possibilities of future expansion also good; but, he commented, an inadequate job had been done in training the Chinese in the service facilities and in basic principles of industrial engineering. He undertakes a constant review of the schedule and noted that the Chinese program manager had some slight difficulty

Legend
Located Taiwan
Located Philippines
Located U.K.

Job Title	Source	1972			1973												1974						
		Oct.	Nov.	Dec.	Jan.	Feb.	Mar.	Apr.	May	June	July	Aug.	Sept.	Oct.	Nov.	Dec.	Jan.	Feb.	Mar.	Apr.	May	June	July	Aug.	Sept.

Foreign Assistance | Manufacturing & Supply

Job Title	Source
Supply & Planning Mgr.	Aust.
Vendor Dev. Specialist N-1	Aust.
Vendor Dev. Specialist N-2	Aust.
Prodn. Control Specialist	Aust.
Fiera Compt. Manuf. Scheduler	F.P.I.
Escort Cortina BB Specialist	Aust.
Escort Process Engineer	F.O.B.
Program Manager	Aust.
S.Q.A. Engineer	Aust.
Program Admin. Plant Layout & Material Handling Engr.	Aust.
Material Handling Specialist	Aust.
Foundry Specialist	Aust.
Plant Engineering Specialist	Aust.

Annotations on chart: Extension, Actual, Original Timing, Extension 23, Extension

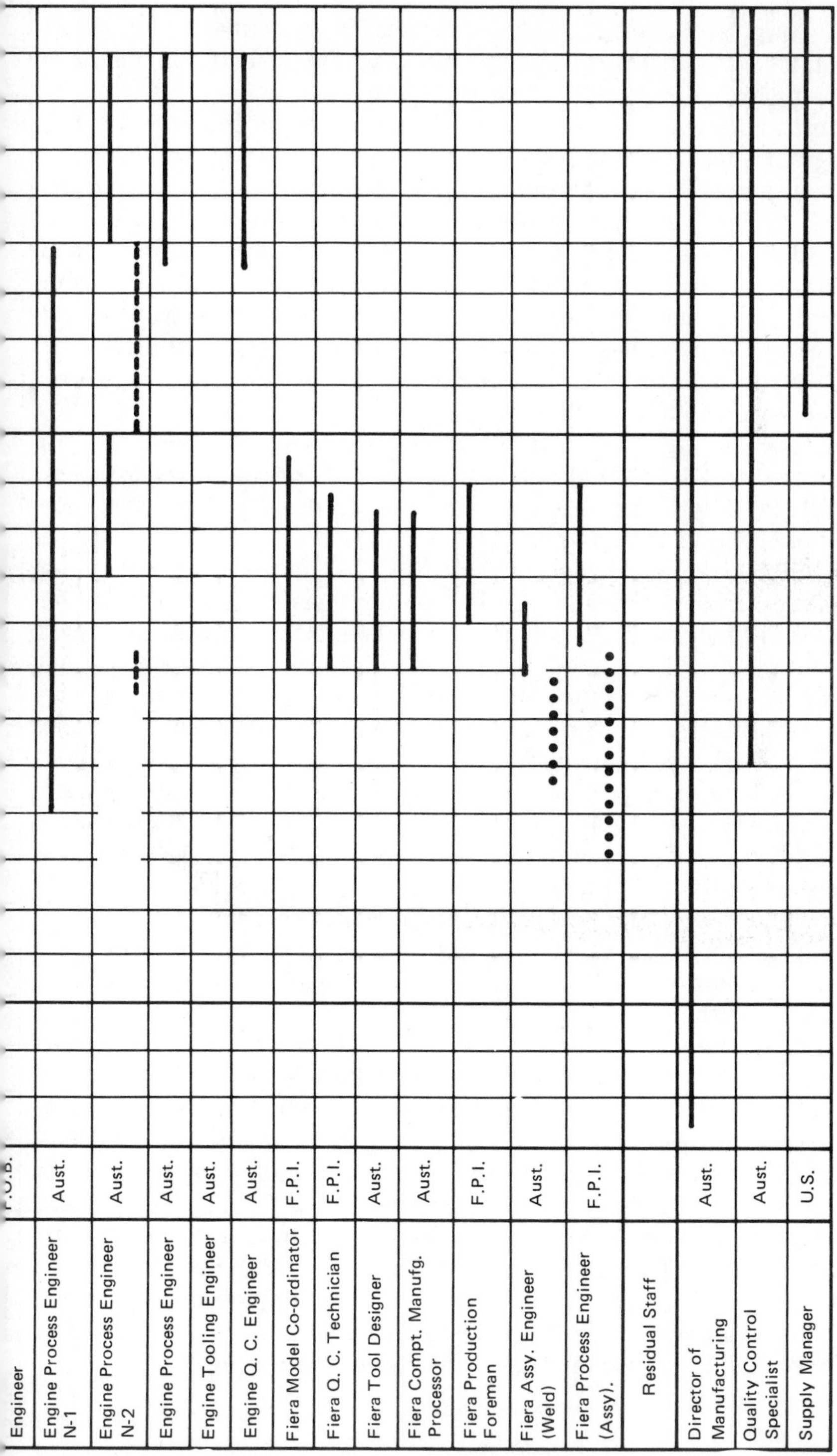

Engineer	T.O.B.
Engine Process Engineer N-1	Aust.
Engine Process Engineer N-2	Aust.
Engine Process Engineer	Aust.
Engine Tooling Engineer	Aust.
Engine Q. C. Engineer	Aust.
Fiera Model Co-ordinator	F.P.I.
Fiera Q. C. Technician	F.P.I.
Fiera Tool Designer	Aust.
Fiera Compt. Manufg. Processor	Aust.
Fiera Production Foreman	F.P.I.
Fiera Assy. Engineer (Weld)	Aust.
Fiera Process Engineer (Assy).	F.P.I.
Residual Staff	
Director of Manufacturing	Aust.
Quality Control Specialist	Aust.
Supply Manager	U.S.

Figure 2-4. Foreign Assignments to Ford–Lio Ho, 1973–1974.

in leaning hard enough on foreign service personnel (who might not take such direction readily), so that the schedule might have slipped a bit.

Visits initiated by FASPAC are not charged to the affiliate; but charges are made for long-term assistance requested by the affiliate. Such assignments can be quite costly; for example, a man and his family in Indonesia for twelve months costs $100,000. The manufacturing staff at FASPAC is composed of eighteen engineers with all-round capabilities; fifteen are from Ford–Australia, two from Ford–Britain, and one from a local supplier in Australia. The cost of assistance from the British company is included in the price of CKD units; but if foreign service employees (assigned over a year or more) were requested, they would be charged separately. The program engineering preactivation and launch costs for Ford–Lio Ho amounted to $1,000,000 plus the costs of preprogram development at FASPAC. Of this total, the cost of providing foreign service personnel was in the order of $750,000.

FASPAC is called upon frequently to short-circuit delays in schedules. For example, when British suppliers went to a three-day week and delivery of equipment was delayed, the Chinese managers did not see how to proceed without this piece being in place. FASPAC engineers were able to show them how to proceed anyway. Similarly, FASPAC officials are needed to help break the bottlenecks with the government over local-content fulfillment and import procedures.

Since Taiwan is part of a production-complementation program for the region, FASPAC officials are needed to audit the performance of each affiliate and to smoothe schedules by keeping peace when necessary, for the key affiliates are interested mainly in themselves and their own performance—not that of others. Dovetailing of expansion plans, therefore, falls to FASPAC, which carries the solutions from one affiliate to another and explains to all how they must fit together.

Once technical expertise is successfully transferred to one affiliate, it can become a source of similar know-how for others. This is especially useful in the complementation program. For example, the Fiera was instituted in the Philippines with 100 percent guidance from Ford-Australia and FASPAC; in turn, the Philippine engineers provided half the assistance to Thailand and 75 percent of that needed by Taiwan. In Indonesia, the Cortina was introduced with assistance from Singapore and the Philippine affiliates. In the development of the next model of the Fiera, Philippine engineers have been brought to FASPAC to be part of the development team; then they will instruct the other affiliates in the region.

FASPAC is responsible for product development within the region, for the smaller affiliates (excluding Australia) do not have this capability. An "Overseas Engineering" group that has been formed within Ford-Australia is capable of drawing on worldwide company expertise for all developing countries. It can develop vehicle design parameters and details and then build up local engineering capabilities to make appropriate modifications. With FASPAC's inception in 1968, a regional group was formed aiming at a still lower level of vehicles than conventional models. The group was formed out of local talent, by drawing on some 50 engineers from the 300 available to Ford-Australia; talents were found that would expand the program capability much broader than needed in Australia alone and suitable to the region. The original Philippines Fiera was designed in Australia by the group; however, tooling for the Taiwan/ Thailand programs was built in the Philippines.

Finally, FASPAC provides continuing training through the programs of its Ford Marketing Institute, which is directed at dealer and service problems. In 1973, 67 separate courses were offered throughout the region and were attended by 1,224 persons from dealers and Ford affiliates.

This description shows an organization oriented to a continuing flow of technology into and throughout the region. The high priority given to these transfers is implemented by assigning responsibilities to key individuals who will be judged in their own performance according to how well they instruct others, how effectively they train their replacements, and how productive the affiliates in the region become in comparison with standards elsewhere.

Chapter 3

Ford-South Africa

In 1973, Ford celebrated its fiftieth anniversary in South Africa. The first Ford assembled in South Africa was the Model T, which came off the line at Port Elizabeth in early 1924. The production line was a small-scale version of the Detroit factory; this was the first technical assistance to South Africa by Ford. In 1940, 5,929 vehicles were assembled from parts imported from Canada. During the War, Ford South Africa devoted all of its capacity to military vehicles; afterwards, Ford expanded its operations, moved into a new plant, and raised production to 12,000 vehicles in 1948. Since then, Ford-SA has shifted its supply source from Canada to a wide range of products from Ford lines over the world and builds cars with bodies from one source, engines from another, and transmissions from another. As a result of expansion, Ford's investment was raised to over $110 million in its several plants in Port Elizabeth, which employ some 6,000 people and produce over 250 cars and commercial vehicles per day plus 15 tractors.

As late as 1962, 87.5 percent of the average car produced in South Africa was imported. In the early 1960s the South African government instituted a five-step local-count program intended to require local manufacturing of 90 percent of the weight of each vehicle. Phase-I of the local-content program required 42 percent by weight to be South African components by 1964. This began the process of local investment in manufacturing, which was accelerated by the requirement of Phase II to raise local content to 62 percent by the end of 1969 and Phase-III's requirement of 66 percent by the end of 1976. (Phase-IV is a two-year hiatus preparatory to Phase V, which is supposed to raise local content to 90 percent or more.)

These requirements brought local production of wheels, tires, hubs, brake discs, rear axles, springs, batteries, seat springs, interior trim and sound deadener, carpets, paints and sealers, glass, cylinder blocks and heads, exhaust systems, radiators, drive shafts, clutch housing, fuel tanks, floor plans, and lights. Save for the production of engines, Ford's decision was to obtain these from local suppliers and to commence a large-scale "buy-out" program in which it helped to develop local industrial capacity.

Despite the large number of companies involved, there are few duplications among the suppliers—that is, almost every product is supplied by a single company, without competitors. This fact reduces flexibility in supply and raises costs of production for Ford. This quasi-monopoly situation is tacitly supported by the South African government, which prefers further import substitution, expanding local content, than proliferation of suppliers of a single product, even if the result is higher costs—its objective is self-sufficiency. It is pursuing this objective simply by raising the percentage requirement on local content—not by dictating the items that must be purchased locally; nor has the government pressed for a laboratory for auto design or testing. Rather, it has been concerned with production capability. If it presses for 90 percent local content, however, the major new items would be body stampings and transmissions, which are both quite expensive on a small scale and based on high-level technology.

The latest major development has been the establishment in 1973 of the Struandale Assembly Plant for the manufacture of Cortina. Whereas the prior plants assembled several models—German Granada, British Cortina and Escort, and the Australian Fairlane and Ranchero (plus trucks from the United States, Britain, and Australia)—the new plant assembles only the Cortina, including a Cortina pickup truck, designed by South Africa to meet local competition. This plant required a new investment of over $12 million, and it will eventually employ 800 Africans and 250 whites.

Ford's competition in South Africa comes from seventeen auto companies and nineteen that make commercial vehicles. Total sales in South Africa of autos were projected at 225,000 in 1974, plus 117,000 commercial vehicles. Ford ranks second in autos and second in commercial vehicles, but first overall. A total of 44 auto models were produced including those of Ford, plus over 240 models of commercial vehicles. The latter do not have any local-content requirements, such as apply to automobiles. Also there is an "assembled category" for cars that permits large models (Fairlane) to be imported completely; this category ceases in 1976, at which time these models will simply disappear, given the high duties to be imposed. In addition, excise taxes are relatively high on autos and are not applied to commercial vehicles and trucks; therefore, the pick-up is quite popular and has a higher demand relative to other markets.

The future growth of Ford-SA depends partly on governmental local-content requirements, the extent to which other companies can stand the pressure of local manufacture, and the rise in income of non-Europeans in the country. Demand by Europeans is already at the level of 35 cars per 100 population—near the advanced-country plateau of 40 per 100—but consumption by blacks is only one-half vehicle per 100 population. Whatever direction the company takes in its growth, it will be greatly reliant on technical assistance from the parent company and other affiliates of Ford, as the following account demonstrates.

TECHNOLOGY FLOWS: MANUFACTURING AND CONSTRUCTION

Ford-SA is a part of the Latin American Group within Ford's International Automotive Operations. This tie—rather than to Ford-Europe or Ford-Asia-Pacific—is partly historical accident and partly a result of geography. Formerly, Ford–Asia-Pacific, Inc., and the Latin American Group were in a combined group, equal in status to Ford-Europe, Inc. At that time, South Africa was in the Latin American subdivision because of its propinquity to South America and the ease of travel between Johannesburg and Rio. When the subunits were separated, no decision was taken to put South Africa in the European or Asia-Pacific region, despite the fact that the models produced in South Africa originate in those two regions. Consequently, South Africa has no administrative ties to the regional headquarters (Europe or Asia-Pacific) of the companies with which it does most of its interaffiliate business. But it is encouraged to seek assistance from either of these regional headquarters, and does so.

Ford–South Africa has ties in several directions throughout Ford operations worldwide. It was historically a part of Ford-Canada, when that company was 40 percent owned by Canadians and was the management center for the Commonwealth affiliates of Ford. Consequently, it still buys components (KD) from Canada, but since the closer integration of Ford-Canada with U.S. operations subsequent to the United States–Canadian Auto Agreement, such purchases go through Ford-US while shipments come from Windsor.

Requests for technical assistance are channeled through the technical service staff in the Latin American Group to the central staff at World Headquarters. This staff selects the right man from anywhere in the world and requests his services through management channels by using letter, Telex, or phone depending on the urgency. The company asked to detail the expert may demur on grounds that he is needed where he is, and central staff then has to decide whether to push harder or move on to the next man on its list.

On problems arising from the use of components from Britain or Germany, South African engineers go directly to their counterparts in these companies; if a dispute arises over responsibility for any difficulty and liability for the costs of adjustment, Ford-Europe is brought into the discussion. If South Africa has made a modification in the design similar to one adopted by Australia, it might go directly there for assistance in the event of a problem. And, in the case of assistance needed by a supplier, it could go directly to a similar supplier company serving another Ford affiliate elsewhere to obtain help.

Since 1964, two years after expanded manufacturing was generated to meet local contents requirements, Ford-Dearborn has detailed 53 different experts to South Africa with a wide variety of capabilities. Three others were sent from Britain, three from Canada, two from Germany, and one from Australia.

The major transfers of technology to the South African affiliate have related to assembly procedures, machining and assembly of the engine, and the construction of the two new plants in the 1960s and 1970s. The list of individuals and their positions in South Africa is shown in Table 3-1 along with the duration of their assignment.

Assembly

The assembly operations at the original plant at Neave had grown with sales over the decades of the 1930s and 1940s and jumped after World War II. The Neave plant is the largest at Ford-SA with a floor space of 75,300 square meters; it produces 270 units per day (two shifts) and employs over 3,000 workers (60 percent are coloreds, 30 percent whites, and 10 percent blacks). It has received continuing assistance from a variety of sources within the company. For example, a quality control specialist was sent from Dearborn for twelve months and another for ten months; another was sent on assembly problems relating to the body build, metal finishing, and

Table 3-1. Employees Assigned to Ford-South Africa from Other Affiliates, 1964-65

Original Location	Original Position	Position in South Africa	Duration of Assignment
Ford–US	Industrial Engineer	Manufacturing Department Manager	Jan. 64 - June 65
Ford–US	Superintendent—Production	Department Manager—Production Control	Feb. 64 - Sep. 66
Ford–US	Product Equipment Designer A	Machine Tool Designer—A	Mar. 64 - Aug. 64
Ford–US	Product Development Engineer Senior	Supervisor Plant Quality Control	Mar. 64 - Jan. 65
Ford–US	Service Manager	General Foreman—Manufacturing	Mar. 64 - June 65
Ford–US	General Foreman	Department Manager—Quality Control	Mar. 64 - Mar. 66
Ford–US	Superintendent—Production	Production Manager	Mar. 64 - Sep. 65
Ford–US	Special Assignment Product Simplification Study Group	Special Assignment Pre-Production Control	Mar. 64 - Feb. 65
Ford–US	Manufacturing Process Engineer Senior	Process Engineer Supervisor	Apr. 64 - Sep. 65
Ford–US	Quality Control Specialist	Quality Control Specialist	Apr. 64 - Mar. 65
Ford–US	Production Foreman	Mechanical Engineer	May 64 - July 65
Ford–US	Tool Requirements Analyst	Tool & Equipment Analyst—A	May 64 - May 66
Ford–US	Production Foreman	Manufacturing Foreman	May 64 - Aug. 65
Ford–US	Manufacturing Process Engineer—B	Manufacturing Process Engineer Senior	May 64 - July 65
Ford–US	Supervisor Divisional Accounting	Supervisor Treasury	May 64 - Apr. 65
Ford–US	Principal Design Engineer	Chief Engineer	May 64 - Mar. 71
Ford–US	Administration Supervisor Product Engineering	Department Manager Administration Product Engineering	June 64 - Mar. 65
Ford–US	Engine Plant Superintendent	Engine Plant Manager	July 64 - May 66
Ford–US	Manager Profit Analysis	Secretary/Treasurer	July 64 - Feb. 67
Ford–US	Industrial Relations Manager	Industrial Relations Manager	July 64 - Aug. 66
Ford–US	Unit Supervisor—Purchasing	Special Assignment—Purchasing	July 64 - May 65
Ford–US	Purchasing Agent	Interim Special Assignment—Purchasing	Aug. 64 - Mar. 65
Ford–US	Staff Financial Analyst	Secretary/Treasurer	Aug. 64 - Mar. 70

Table 3-1   continued

Original Location	Original Position	Position in South Africa	Duration of Assignment
Ford–US	Accounting Procedures & Reports Analyst	Supervisor—Treasury	Sep. 64 - Nov. 64
Ford–US	Manufacturing Engineering Department Manager	Manufacturing Department Manager	Sep. 64 - Feb. 66
Ford–US	Manager Vehicle Concepts Department	Product Engineering Manager	Oct. 64 - July 67
Ford–US	Financial Analyst A	Manager Investment Analysis	Sep. 64 - Mar. 66
Ford–US	Product Development Engineer A	Section Supervisor Product Engineering	Oct. 64 - Jan. 69
Ford–US	General Foreman	General Foreman	Dec. 64 - Feb. 66
Ford–US	Production Test Engineer A	Section Supervisor Engineering	Jan. 65 - Aug. 67
Ford–US	Supervisor Programming & Release Product Engineering	Department Manager Administration Product Engineering	Jan. 65 - Feb. 66
Ford–US	Sale Engineer	Principal Design Engineer	June 65 - Nov. 66
Ford–US	Contract Annotment Research Analyst	Salaried Personnel Specialist	
Ford–US	Manufacturing Specialist	Manufacturing Process Engineer—A	Feb. 65 - Apr. 66
Ford–US	Supervisor Overseas Scheduling	Product Programming & Timing Manager	June 65 - Mar. 66
Ford–US	Programming Timing Co-Ordinator	Department Manager—Product Engineering	Jan. 66 - Sep. 66
Ford–US	Executive Engineer	Product Development Manager	Feb. 66 - July 67
Ford–US	Dealer Finance Manager	Dealer Finance Manager	June 67 - Feb. 70
Ford–US	Pre-Programming Engineering Manager	Product Development Manager	Aug. 68 - Jan. 72
Ford–US	Finance Manager	Secretary/Treasurer	Feb. 70 - Sep. 72
Ford–US	Senior Engineer	Resident Engineer	Mar. 70 - Oct. 72
Ford–US		Ford Credit Manager	May 71 - Dec. 72
Ford–US	Operations Controller	Controller—Finance	Dec. 71 - Jan. 74
Ford–US	Branch Manager	Dealer Finance Manager	Jan. 73 - July 74
Ford–US	Executive Director	Managing Director	June 70 - To Date
Ford–US	Product Planning Manager	Product Planning Manager	Jan. 73 - To Date
Ford–US	Export Supply Manager	Supply Manager	Jan. 73 - To Date
Ford–US	Assistant Distribution Manager	Vehicle Sales Manager	Jan. 73 - To Date
			Feb. 73 - To Date

Table 3-1 continued

Original Location	Original Position	Position in South Africa	Duration of Assignment
Ford–US	Secion Supervisor	Marketing Staff Manager	Sep. 73 - To Date
Ford–US	Body & Paint Specialist	Production Operations Manager	Nov. 73 - To Date
Ford–US	Department Manager	Quality Control Representative	Jan. 74 - To Date
Ford–US	Branch Manager—AA	Field Operations Manager Ford Credit	Feb. 74 - To Date
Ford–US	Financial Analyst—A	Treasury Manager	Nov. 72 - To Date
Ford–Canada	Integration Planning Manager	Forward Planning & Programming Manager	Jan. 64 - Aug. 64
Ford–Canada	Supplier Quality Assurance Representative	Process Die Design Construction & Stamping Co-Ordinator	Oct. 64 - Apr. 65
Ford–Canada	Product Timing Manager	Program Timing Manager	Feb. 65 - June 65
Ford–Australia	Market Representation Supervisor	Market Representation Manager—Tractors	Mar. 70 - July 71
Ford–Germany	Product Development Specialist	Product Development Specialist	Jan. 68 - Aug. 68
Ford–Germany	Product Development Specialist	Product Development Specialist	Jan. 68 - Aug. 68
Ford–Britain	Product Design Engineer—A	Purchase Analysis Specialist	Mar. 64 - Mar. 65
Ford–Britain	Dealer Finance Manager	Dealer Finance Manager	Mar. 66 - Aug. 68
Ford–Britain	Design Engineer	Principal Engine Design Engineer	June 65 - Dec. 68

paint. A problem of blistering under the paint brought an engineer from the U.S. manufacturing staff within 48 hours.

The requirement of local content in the early 1960s meant that Ford-SA engineers had to understand manufacturing—not just assembly—which it had not learned before. In 1964 a six-man team that was sent out from Dearborn's supply staff to educate the SA engineers in materials and supply procedures stayed for over a year. This team checked on the capabilities of different suppliers and helped to select companies that could produce to quality requirements and on schedule. To help set up the supply program, SA sent one engineer to Dearborn on purchasing procedures and techniques, and a Canadian purchasing agent was sent in return to Port Elizabeth; during a three-year stay, he established a computer program on parts-release and set up a traffic pattern on parts handling. Still later, the vice president of supply and purchasing came out from Dearborn with new ideas on storage and purchasing. In 1970 and again in 1971, two officials came out from Dearborn to set up and advise on the "Best Ideas Production Materials Information System" (BIPMIS), drawn from Ford experience around the world. To follow up, Ford-SA sent five men to Ford-Argentina to see how they used this system.

At the initiation of assembly of the Fairlane and Ranchero—models produced in Australia—there was an exchange of personnel between SA and Australia, followed by others a year later. Whenever new models were introduced by the European suppliers of the basic components, one or two manufacturing engineers were sent out, as well as a process engineer to facilitate training of SA personnel.

Technical assistance has not flowed only into Ford-SA, for it has also helped others coming into operations similar to its own. Ford-SA engineers have been detailed to Rhodesia, the Philippines, New Zealand, Taiwan, Thailand, to the Asia-Pacific staff—for as long as two years—and some have gone on extended training assignments, as shown in Table 3-2. These assignments provide new insights to the personnel, for in pursuing new problems they come up with different answers than at home; or, in checking on *why* a given procedure is followed, the answer obtained on the shop floor is often different from that given by engineers (the procedures actually followed are often different from those specified in drawings). They find that to transfer technology appropriately there is no substitute for face-to-face discussion of a particular problem as it has arisen.

Supplementing these visits is the "feedback" program, which provides for regular exchanges of information among affiliates on

Table 3-2. Employees of Ford–South Africa Transferred to Other Ford Affiliates, 1960-1974

Original Position	Position on Assignment	Duration of Assignment
Supervisor Manufacturing Engineering	Special Assignment Manufacturing Ford–Mexico	Jan. 60 - Oct. 60
Supervisor Production Control	Manufacturing Engineering Manager Ford–Rhodesia	Dec. 60 - Apr. 67
Quality Control Manager	Production Manager Ford–New Zealand	May 62 - Mar. 63
Production Manager	General Manufacturing Manager Ford–Malaysia	Jan. 63 - Dec. 65
Tractor & Implement Manager	Managing Director Ford–Malaysia	Jan. 64 - Sep. 67
General Tractor Manager	Executive Director—Tractors Ford–UK	Sep. 68 - To Date*
Salaried Personnel Manager	Personnel & Administration Manager Ford–Thailand	Nov. 70 - Nov. 72
Material Control Manager	Material Control Manager Ford–New Zealand	Feb. 71 - Aug. 71
Assistant Controller	Profit Analysis Manager Ford–Australia	Apr. 72 - June 74
Market Representation Manager—Tractors	Market Representation & Business Management Manager—Tractors Ford–US	July 72 - To Date*
Purchasing Agent	Supply Manager Ford–Phillipines	Feb. 73 - To Date*
Financial Budget Analyst	Controller Ford–Thailand	Mar. 73 - To Date*
Product Design Engineer	Product Engineering Manager Ford–Thailand	July 73 - To Date*
Car Planning Manager	Car Product Programs Manager Ford–US	Dec. 73 - To Date*

*As of Sept. 74.

ideas others have developed and used. Ford–SA has submitted one idea each quarter, as requested, including some related to a "centralized robbing facility," a "quick change adaptor," and "seam welding of fuel tanks." It has responded to enquiries from Ford–Philippines on switch-off motors and inspection travel cards, but does not know if any of its ideas have been adopted by others. And it has adopted several of the feedback suggestions of others—such as the "clutch pedal setting guage," "Goodway binding strap cutter," "supplier-quality-assurance wiring supplier checklist," and a "frame expender"—and has several other items under active consideration.

Engine Plant
A major stimulus to technical transfer was the introduction of local-content requirements in the early 1960s, which gave rise to the establishment of the engine plant. The engine plant has 16,500 square meters of space, produces 300 units per day, and employs over 450 workers (60 percent blacks and 40 percent whites). The determination to establish the engine plant was a joint decision among South Africa, the Latin American Group, and Dearborn officials. The choice of the engine to be produced was made in conjunction also with the Ford–UK, which backstops the operation in South Africa.

The entire plant construction, layout, and processes were developed by the manufacturing staff in Dearborn out of its worldwide experience. A resident engineer was sent from Canada by Dearborn to oversee the construction of the building. The provision of such a "one-shot" expert cuts the costs of advice and increases efficiency not only of the plant but also of later technical assistance because of the similarity in operations. In addition, a comprehensive exchange of technology took place over the two years 1962–64; some 40 people were selected out of offers from manufacturing units all over the world (only those beginning in 1964 are recorded in Table 3-1). Of these, 35 came from various units within the United States, 1 from Britain, 1 from Australia, and 2 from Canada. They brought differing levels of technical ability and varied expertise and stayed between 12 and 30 months.

Dearborn decided that certain equipment and processes used in the British plants would be best for South Africa. The equipment was ordered in the United Kingdom and installed by the manufacturers. Local engineers and labor were hired—raw and untrained in engines—to begin learning the jobs of the 40 expatriates. In anticipation, the present product development manager had been sent to Canada for two years, then to the United States to work on the Ford team preparing the new plant project, and then back to South Africa with the team of 40.

The expatriates held the positions of project manager, plant manager, manufacturing engineers, plant and equipment maintenance, production maintenance, materials control, and quality-control engineers. The replacement for the plant manager was sent to the United States for training; the plant engineer and manufacturing engineer were sent to Britain to visit the equipment manufacturer and all division heads to learn their jobs, while others took understudy positions in Port Elizabeth. Almost all of these replacements were promoted from within the company, from the Neave Assembly Plant.

The major complexity in operations at the engine plant was that two different engines would be produced with the same machinery and on the same production line. These were the Kent (in-line 4 cylinder) and the Essex (V-4 and V-6) for the Escort, Cortina, Capri, and Zephyr Zodiac (later the Granada).

Technology transfers encompassed, therefore, building construction, (including *all* amenities and support facilities), plant layout, processes, equipment, and model selection prior to start up. After start-up, the training of replacements was completed within a couple of years, and South Africans took over all positions in the plant.

Technical assistance continued, however, on a variety of activities, such as materials handling. Given the number of materials imported, South Africa has one man checking full-time on procedures for unpacking and handling the nested components and parts; he is in constant contact with the British and other suppliers and attempts to improve methods of nesting and protection of supplies so as to reduce costs of freight and damages to the components. These exchanges of information have saved between $200,000 and $300,000 each year.

The engine plant receives parts from Britain and machines the block, the engine head, bearing caps, fly wheel, and the crank-shaft pully. (It also machines brake drums and discs, because of the lack of acceptable local supplies, while receiving technical assistance from both the United States and Britain on processes and quality assurance.) These and other processes are aided by the distribution of a booklet on manufacturing developments put out each quarter by the United States and another by Ford–Germany on basic manufacturing processes; these booklets are directed to affiliates that are manufacturing an entire line, but they are still helpful.

Despite the fact that the engines come from Britain, the U.K. company is not a "mother-company" to South Africa. Rather, the SA affiliate is encouraged to search worldwide for information by going through channels at headquarters, which may then assign anyone from over the world who has relevant experience. This procedure for visits is supplemented only infrequently by contacts by phone or Telex when someone knows that specific experience can be delivered at long distance.

A recent technology transfer to the engine plant was in connection with preparations to machine the crankshaft. In order to increase local content, another part with substantial weight was needed; the crankshaft seemed to offer the best possibilities. (Formerly the total engine qualified for local content, despite the import of components that could not be obtained locally; with the rise of local capabilities, more pieces had to be obtained locally; the crankshaft was first on the list.) Eight SA engineers went to the United States, Britain, and Germany for one to three months for observation of the processes.

Ford-Australia had begun this same process earlier, so its specifications were drawn upon, and it was asked to check work done by SA engineers in drawing up recommendations for Dearborn. Since machinery was to be purchased in Britain, two engineers went to examine the alternative equipment and make final recommendations on purchases. Although SA did not need the same extensive assistance that it did in setting up the engine plant, and therefore did not bring in Dearborn specialists, Dearborn's manufacturing staff did check all aspects of the production layout, specifications, machinery, and so forth. This assistance will maintain quality control and prevent costly errors in the switchover.

Cortina Assembly

Because of the demand for the Cotrina model, which had been upgraded by addition of a 6-cylinder engine, it was decided to set up a separate assembly plant for it at Straundale. By mid-1974 it had been in production only nine months but was up to 80 percent of capacity. The plant has 36,200 square meters of space; it produces 100 units per day, and employs over 520 workers, of which 75 percent are blacks and the rest whites.

Technical assistance in the set-up of this plant was also received from Dearborn, principally in the detailed review of all plans. SA started with the capacity of plant in mind, plus later expansion, and added information and criteria developed out of past experience and its numerous visits to Ford facilities around the world. The present general manufacturing manager of Ford-SA had been sent to Ford-Rhodesia in 1960 to help set up the manufacturing operations there and stayed four years; the present plant and manufacturing engineering staff manager had gone with him but stayed seven years, then had gone to the Philippines for four years to set up the plant there. This experience in staffing a plant with blacks in Rhodesia was invaluable to these managers, since Straundale is staffed almost wholly with blacks, including 7 of 47 foremen. The result of this technology transfer through foreign experience was to achieve an exceptionally smooth start-up of the Cortina plant.

Dearborn sent one plant engineer out for six weeks on site selection, service buildings related to future expansion, and long-range planning. Dearborn's manufacturing staff had already developed for general use, work standards, layout, conveyers, manufacturing lines, stock storage facilities, and so forth, so that SA merely had to "put them together" in the appropriate package for Cortina and the size production desired. On a visit to Dearborn to explain the project, the general manager of Ford-SA and an engineer discussed problems of process engineering, lengths of lines and conveyers, materials handling equipment, stocking and stock layout.

SA officials estimated that if they had had to buy this information outside of the company, the plant would have been delayed at least a year and would have required 360 man-months or a *penalty* of 240 man-months over the time actually required, and would have cost an additional $300,000 or more. Therefore, the cost-advantage of having in-house technical assistance was a year's time and some $300,000 in cost savings in the mere construction phase. Considerable additional assistance was obtained from manufacturing staff, from the security and fire protection staffs in Dearborn, and on construction of a computer facility for stock control. To learn how to run the "parts-release system" better and to improve the "dealer-order-taking and vehicle-delivery" system, three or four officials were sent from SA to Latin American affiliates for examination of their computer programming. Since then, SA has modified the system and sent back its improvements to Latin American officials, who have adopted some of the changes.

The most significant technical assistance came in the adoption of the electro-dip paint process, which eliminates undercoating and is much faster and cheaper. The SA manufacturing manager had seen this equipment in operation in Australia and sought it for Straundale. The paint tank incorporates the "ultra-filtrate" system, which gives purer quality and saves paint. Australia helped by designing the requirements for SA and making modifications from its own experience; the U.S. manufacturing staff checked these out. Australia also watched over the construction of some of the equipment by Siemens.

To use the system, SA also needed a paint supplier. Ford–US advised SA to talk with Glidden's licensee in South Africa because of Glidden's association with Ford–US; since the licensee was not making this type of paint in South Africa (though it was making electro-dip paints for VW and Toyota), it had to obtain new technology from Glidden–US. A design team was formed including an engineer each from the U.S. manufacturing staff, from Glidden, Siemens, and the British parent of the Drysys company in South Africa that was to make the equipment. An Australian brought out the designs and stayed two months to help finalize them. The contract for the power supply was let to Siemens even before the rest of equipment was designed because of the leadtime required; very close liaison was maintained between Siemens and the other contractors. The responsibility for the system was Drysys', with Ford–US and Glidden interested "bystanders," but several engineers were in constant touch on the progress.

A paint engineer from SA was sent to European affiliates—Dagenham in Britain, Genk in Belgium, Saar-Louis, and Cologne, Germany—to learn about the electro-dip system and other aspects of production, in preparation for his later promotion to manufacturing engineer. He also visited the Carrier Corporation, which is the parent company of Drysys (in South Africa). Another engineer went to Dearborn to present the design-contract package to the manufacturing staff; they changed some specifications and equipment and the layout in order to get the equipment into proper place and working most efficiently.

Ford developed the entire system originally, including the machinery and paints, and licenses it to other auto companies and paint producers. The Drysys equipment can be used with other paints, as was being done already in South Africa, but to produce the Ford paint, both Glidden and its SA licensee pay a royalty to Ford–US on each gallon supplied.

Training and Promotion
One of the principal purposes of technology transfer, of course, is to train workers so that they can produce more efficiently and raise their own standard of living as well as achieve some self-satisfaction in a job well done—in line with Henry Ford's policy statement that "Our role is to contribute more to the quality of life than mere quantities of goods."

The company maintains a "promotion-from-within" policy and therefore encourages all workers to upgrade their skills and capabilities; in addition, it places on all supervisors the responsibility for providing subordinates with opportunities for advancement. Advancement comes through work experience and training.

Upon entrance into the company, workers are given an indoctrination program to familiarize them with work conditions (when to come to work, tea breaks, why to stay on the line), personnel policies, and the education and training programs offered. The education programs include classes in basic education—up to level of Standard 4 (sixth grade)—which are given on company premises after work.

The basic education classes provide reading and writing skills for workers with no education or literacy; these courses bring the workers up to fourth-grade levels in three or four months and prepare them to go to night school. General education classes through high school levels are given in night schools; and a substantial number attend. Classes are offered to any employee with a Standard 6 who wishes to study further in a technical direction. Training programs are offered to improve operator skills in such tasks as spot welding, mechanical repair work, and so forth; courses in the fundamentals of supervision for group leaders; courses to train foremen and supervisors, including a two-year course on technical skills needed by first-line supervisors. All expenses are paid by the company as are all expenses for post-high school education if the employee meets the completion standards of the courses taken, but the number participating is disappointingly low. Figure 3-1 shows all programs in 1974.

For university training, Ford has established scholarships (bursaries) for students selected by university officials; it has also donated equipment to technical high schools and is helping build a science and mechanics lab attached to a high school for blacks in Port Elizabeth. The company also invites high school principals to visit the plant so they can explain to students different opportunities for employment.

Through these training programs, blacks and coloreds have been promoted into foremen positions, and it is expected that they will go still higher, reflecting their capabilities and the increasing scarcity of European workers. At present the same wage is paid for a given job, regardless of who does it—black, white, or colored. Ford's minimum wage is higher than the Household Subsistence Level (formerly Poverty Datum Line) as published by the University of Port Elizabeth. The average nonwhite hourly income, which includes the value of fringe benefits and normal paid overtime, exceeds the Household Effective Level. The scales for the middle grades between 1 and 12 are shown as of October 31, 1974:

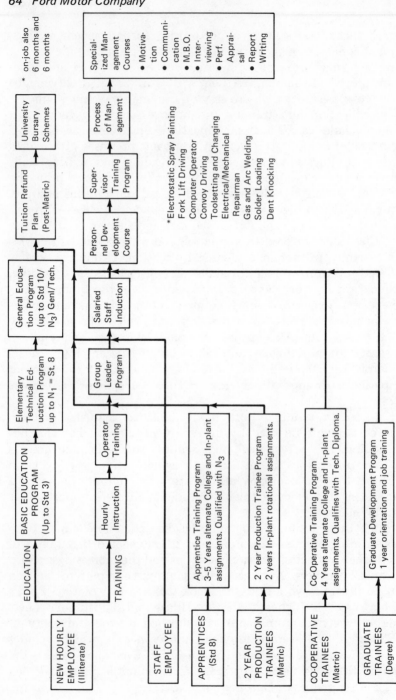

Source: Prepared by R. A. Waddington, Education and Training Department, Ford Motor Company.

Grade	Coloreds	Africans	Whites	Average Hourly Rate (Rands)	Monthly, Including Benefits	U.S. $ (Equivalent)
5	226	75	1	0.75	R 180.53	270
6	243	80	4	0.81	R 193.87	290
9	193	64	244	1.16	R 288.16	432
10	11	0	158	1.41	R 354.33	531

The number of workers receiving training has increased rapidly in the past two years as a result of the turnover due to pirating of employees by other auto companies recently entering the country; in addition, Ford is now reaching the time when a number of floor supervisory staff are retiring with 30 and 40 years service. And, of course, the introduction of the engine plant and the Cortina assembly plant required extensive programs for all levels of new and old workers.

At the establishment of the engine plant, both semi-skilled and unskilled workers were brought in from the outside, and blacks were moved into formerly white jobs. They were hired in groups of 5 or 6 workers over a period of 18 months for a total of 150 per year. They had no knowledge of engineering and were put into tasks that were deskilled initially to fit the abilities of the workers; but each was given training that helped him understand what his function had to do with the whole engine. The initial instruction was followed by on-the-job training both on an individual basis and in groups. This training includes mathematical concepts and measuring instruments, skills in quality control and supervision, use of blueprints, welding, spray painting (takes 6 months), and so forth. Success will permit promotion from apprenticeship to artisan status. These courses last for one year, and a student has a 90 percent chance of promotion afterwards.

At the engine plant, there are 70 workers in the assembly operation, with 4 group leaders without work responsibility of their own (grade 8 in the job scale) who are also blacks; the two foremen are white, but they will eventually be replaced by blacks as these latter get training in handling men. Among those who can achieve a job grade of 8, there are 86 in the engine plant, for they have a Standard 8 education (equal to tenth grade) or better; of these, 20 have a Standard 9 (eleventh grade) and 10 have graduated from high school. A grade-8 job would include tasks such as quality control inspector, storeman, tool changer, and group leader; grades 3–5 are machine operators. Less than 3 percent of the workers are illiterate (only 10 in number), and they are older workers who either have no ability or simply will not make the effort.

With the opening of Straundale, Ford embarked on a large-scale program of hiring blacks. It had not hired blacks previously in production jobs, so it moved coloreds into Straundale to get the operation going. Some newly hired blacks were put into Neave to obtain on-the-job training; they were shifted into Straundale when they were able to produce and help train new employees there. (Once the process was completed, the coloreds were returned to Neave.)

Training of blacks requires repeated oral instructions, for the Bantu languages are repetitive—that is, they build from sentence to sentence by adding a new phrase after repeating the first, and so on until the last sentence gives a full description. If the black worker does not understand, it is less because he cannot than that he was not listening at first and was waiting for the repetition to show the complete thought and its importance. In the main, he is not addicted to reading and cannot visualize from diagrams or verbal description; he must see a problem and solution in operation. While this causes some difficulties, it reduces others, for he is not offended by others having ideas or concepts derived from theory. He has no NIH syndrome, therefore, and willingly learns. He will believe what he is shown and follow instructions. If he later is told that what he has learned is *not* right, he will become upset and not know where to turn for correction; he must be given the new, correct procedure at the same time. If he must make a decision, he wants group reinforcement.

The production methods employed are deskilled with the result that the volumes are not as high as in the United States. For example, the lower cost of labor makes it feasible to use a single-head pneumatic wrench for nuts on wheels, rather than multi-head wrenches. Women are used only in lighter operations in the plant, such as wiring and assembly of instrument panel; however, one wanted to drive a pallet-truck in the plant and became one of the best operators.

Promotion of workers will increase not only through expansion of production but through exemptions from "job reservation" regulations; these exemptions are given to permit expansion, that cannot occur because of the continued tight job market for whites (many of whom hold jobs for which they are not qualified simply because they are European). In order to maintain its policy of "promotion-from-within," Ford–SA had to accept some promotions of individuals to supervisors in the new plants who did not meet leadership requirements. (In one case in the paint shop, a black foreman had to be replaced by a white employee because of a lack of ability.) In the main, when blacks do become grade-8 supervisors, they are as good as their white and colored colleagues. To broaden capabilities and obtain a wider selection of supervisors, Ford put selected workers through special classes that familiarized them with many operations, even if they could not perform them all. All the quality-control technicians are blacks as are half of the inspectors, though none are in final inspection. It requires some three months for training in quality control, but blacks were assigned after only six weeks during the first operation of the Straundale plant.

These efforts at promotion seem to reduce turnover, which occurs mostly within the first three months of employment. With the opening of the Straundale plant, blacks were sought with high school education and several were obtained—even two or three with university training. To get more coloreds, however, Ford–SA has had to take those with poorer preparation, and they do not stay as long. Whites do not stay on as technicians or artisans in such tasks as toolmaking, fitting, turning or electricians; so blacks and colored could rise readily if they had the education and training—but they do not as yet.

The promotion policy includes an appraisal of every individual at specific intervals, and lists are kept as to the abilities of each worker as he is moved about on the production line (e.g., from assembly to machining of blocks, to metal finishing, and so on). The development of multiple skills increases the opportunity for promotion.

Training to become a supervisor requires 18 months; Ford seeks individuals with a high school diploma who are willing to go through new training; once they have completed this program they will become supervisors and inspectors, at relatively high job levels. In addition, courses are required of salaried employees in subjects such as motivation, communication, problem-solving, and management by objectives.

Only whites are employed in the top two hourly job grades of 11 and 12,which cover artisans and technicians; no coloreds or blacks are yet at that level, but they are likely to be promoted because of the scarcity of whites (unemployment for whites and coloreds is less than ½ percent); Consequently, nonwhites have been undergoing training to raise them to grades 11 and 12. The distribution of employees in training programs in 1972 is given in

Table 3-3. No blacks or coloreds were enrolled until 1970, and increased participation for them was encouraged by the statement of the Prime Minister in 1973 that less stringent controls would be exercised over use of nonwhites in the motor industry.

At still higher levels, engineers, continue their training through visits to other affiliates, where they are detailed for short or longer stays. When they return they are slated for promotion. For example, the present manager of product development (one of the top managerial positions) was formerly chief engineer and prior to that manager of the engine plant. To prepare for promotion to chief engineer, he was sent to Britain, Australia, and Germany for over a year. The next chief engineer is now in Taiwan as resident engineer helping that affiliate set up the Kent engine line. Another manager was in Dearborn with the Latin American Group for training on product planning and research and is to be slotted in one other area before returning to South Africa. During the 13 months of July 1973–July 1974, 31 trips were made by Ford-SA officials to the United States and other affiliates; 17 of the trips included a visit to the United States, 5 to South America, 12 to the United Kingdom, 4 to Germany, and one each to France, Australia, and Hong Kong.

Of the over 6,000 employees, more than 3,000 received assistance for some form of education or training outside the company,

Table 3-3. Employee Participation in Ford-South Africa Training Programs, 1972

Program Title	No. of Participants			
	African	Colored	White	Total
Basic Education Program	23	9	0	32
Elementary Technical Education Program	30	54	0	84
Tuition Refund Plan	2	9	83	94
Two-Year Production Trainee Program	0	4	0	4
Apprentice Training Program	0	0	37	37
Workshop Trainee Program	0	8	0	8
Cooperative Training Program	0	0	19	19
Toolsetter Training Course	3	0	0	3
Chargehand Training Program	0	39	30	69
Salaried Staff Induction	2	2	76	80
Personnel Development Course	1	4	92	97
Graduate Development Program	0	0	7	7
Management Training	1	0	20	21
Supplier Management Course	0	0	40	40
Motivation Course	1	0	44	45
Communication Course	1	0	55	56
Total	64	129	503	696

while another 1,000 attended in-company and external courses and seminars in 1973. The costs of these programs for 1973 amounted to over $300,000 and was expected to top $375,000 in 1974.

In addition, Ford has provided training aides, engines, transmissions, and so forth to the Zwelethemba Trade School to help students prepare for auto mechanic jobs; the company has indicated a willingness to hire these graduates. It has also initiated a training program for primarily black mechanics working in the Government Garage in Kingwilliamstown, the capital of Ciskei.

In sum, this review of technology transfers points up the importance of a continuous flow and the changing nature as markets and models change. Without such transfers, the recipient would not produce as efficiently. The effect of cutting off such flows is seen in the experience with TATA trucks in India, based on technology acquired from Mercedes Benz. The result has been static and poor design, high capital costs, and poor (nondynamic) management. The same can be observed in the Standard Motor Car Company of the United Kingdom with its Herald model, which faded from the scene.

ASSISTANCE TO SUPPLIERS

Outside of the manufacturing processes within the company itself, technology is transferred through assistance to suppliers. This procedure was accelerated greatly with the introduction of local-content requirements. Since Ford adopted a policy of "buy-out," rather than manufacture itself, it buys original equipment from 152 suppliers; the annual bill in 1973 was over $45 million, and another $15 million went for service replacement parts or general stores. Of these suppliers, some 80 have affiliations with foreign companies—either as subsidiaries or licensees; those that are licensees frequently have such contracts from more than one licensor abroad. These associations were encouraged by the South African government. Ford, however, has not taken the initiative to bring in a foreign company, though several have inquired about the feasibility of establishing subsidiaries. Ford–SA would not discourage such action, for it would get better service and technical quality; also the foreign company is already up on latest developments and would not require technical assistance from Ford. A local company has a lot of catching up to do, and Ford must go through an expensive prototype program to get it into production.

Direct Aid
Despite the difficulties, Ford has brought many South African companies into the motor industry. For example, the Felt and

Textile Company was encouraged to move into production of molded carpets for autos; GKN was given its boost into drive-shafts by a Ford request; Silverton was brought into radiators; the Lectrolight company was helped by Ford to get into manufacturing of generators in the early 1960s; NCI was encouraged to go into plastics trim; Quintin Hazel was introduced to exhaust systems; Rolex into light stampings; ROMP into bumpers; another company into seat springs and frame assemblies; Pascor into seat-spring manufacture; National Spring into road springs; G. C. Shaves into paints; Duroplastics into plastic seating; Bag Stores (previously in suitcases) into plastic panels; Jackson's (formerly in toilet seats) into steering wheels; and Trek Tools into small stampings.

Both these and already established suppliers have required extensive technical assistance that ranged from manufacturing processes to scheduling, quality control, cost control, managerial advice on organizational structure, purchasing analysis and packaging. Assistance on quality assurance is given to all suppliers, and some management training courses are offered on subjects ranging from industrial relations to purchasing; demand is greater than Ford's ability to supply.

A Supplier Quality Assurance program is used to obtain quality control at the point of parts production, thereby eliminating inspection by Ford on delivery of items to its plants. It is so successful in obtaining fulfillment of standards that Ford's competitors have copied it with their own suppliers. The assistance given by Ford and GM to local suppliers is of considerable advantage to other auto and nonauto producers, who benefit from the results in better quality and scheduling. For example, Ford's specifications on paints and machinery are simply adopted by others who ask suppliers to give them the same thing.

Several Ford engineers are field representatives who visit suppliers by area; they ferret out problems and find solutions for them. Sometimes the problems are so extensive that they require continuous help. In one case, production planning and control problems were so difficult for a seat-spring plant that a Ford man literally ran the operation for a time. In another, when a key manager fell ill, a Ford man was detailed to fill his place, for there was no one else to take over.

A parts release and control system (inventory management) is given to each supplier also, so as to dovetail his production scheduling with delivery of imported parts. To help institute the system, a Ford man will stay a half day or longer with each supplier and then go back when the supplier's personnel changes or when Ford changes its specifications or production planning. And three officials are kept busy constantly checking on suppliers to see whether they are on schedule. Delay is a serious problem as evidenced by the fact that Ford had over 380 vehicles in lots outside the plants awaiting parts at the time of our visit.

Suppliers are also assisted in going abroad to visit similar companies in Europe or the United States, whether they are Ford affiliates or not, though most of the visits are to Ford divisions producing similar items. For example, the foundry company in South Africa sent engineers to the Ford foundries in the United States and Europe.

Assistance to suppliers is both on a regular basis and to meet immediate problems. For example, ROMP was checked out thoroughly on quality control but began to slip; a Ford grade-8 employee was sent to bring it back to scratch, which he did; then he interviewed new personnel to up-grade the technical level of the company. Over several months, several grade-7 and -9 operators were sent to help set up a production line and train operators for the GKN Sanky company, which had imported metal pressing tools but did not know how to set them up or maintain them. In another case, the producer of high tensil (wheel) bolts was having a problem with heat treatment; four days of assistance was required to meet the problem. In another, a vent disc on the Granada was cracking around the edges; Ford got assistance from the United Kingdom and passed it on to the supplier. A licensee of a U.K. company producing locks had a problem of materials control and heat treatment that resulted in the fracture of a main piece of the lock; two Ford technicians (grade 8) and a metallurgist were sent for four days to Capetown; they devised a system of control of material handling, sorting, and process controls, and retrained the heat-treatment supervisor.

Suppliers face many problems in quality control in the metallurgical field because of bad practices of engineers, who have had theoretical training but no practice in their university programs. Ford fills the voids of industrial engineering to some extent; when it cannot succeed—as it couldn't in the case of door handles—it simply has to import.

On packaging, Ford issues specifications that suppliers are to follow, but they could do much better than they do and would achieve significant cost reductions if they followed instructions. In some cases, the specifications for packing are inadequate and Ford has to counsel with the supplier as to what is feasible and desirable.

Assistance is also provided in purchasing and in materials supply; for example, a few years ago, Ford faced a reduction of supply of steel sheets from ISCOR (South African Iron and Steel Corporation).

The shortage was so bad that Ford started shifting its supplies about and sought permission to import more; ISCOR even asked Ford to find imports for it. Ford-Britain came up with a supply, but then ISCOR refused to authorize the imports and said it would have a sufficient supply for all customers. Ford, in desperation, imported 700 to 800 tons anyway; it received an import permit only after the supply was used up or delivered to suppliers. The suppliers could not possibly have supplemented their normal sources because they are small and without worldwide contacts. This shortage was made up only because of Ford's tie-in with foreign suppliers through other Ford affiliates. Also, in materials such as steel, Ford is able to give assistance to a supplier facing deviations in quality of materials by demonstrating what such deviations may mean to the quality of a component or part.

Sometimes in helping a supplier who is a licensee of a foreign company Ford-SA has had to go through another Ford affiliate to get the licensor to act—as in the case of a British company that was also a supplier to Ford-Britain.

Foundry Problem

The most extensive assistance given a single supplier has been Ford's work with Ferrovorm, a foundry sponsored by the Government's Industrial Development Corporation to produce castings for the auto industry. The company has run into serious problems that Ford has tried to eliminate; it will not be able to produce to specifications until 1975 or after, despite heavy pressure by the government to substitute for imported components. In the meantime, it is producing samples for Ford and still running into a variety of problems.

The story begins over two years ago when Ferrovorm ran into a problem in trying to cast the Essex 6 engine block. Ferrovorm had chosen to use a sophisticated German shell-moulding process in casting, despite advice from Ford and GM both that it would be better to use the older green-sand process, which requires less operator skill, even though it is a bit less accurate. Ferrovorm objected that the conventional method required higher capital costs and argued that its engineers could handle the more complex process. To help persuade the Ferrovorm engineers, Ford sponsored visits to Ford foundries in the United States and Europe, but they still chose the German process. A team from Ford-SA, Ferrovorm, and the Ford Foundry in Britain took the German moulds to Britain to try them out in an attempt to anticipate the problems. As a result, Ford had to change its metal specifications to accommodate the process to be adopted by Ferrovorm. Within a year, as a result of poor overall management in Ferrovorm, about half of the foundry's staff left to return to their native countries or to go into private companies; the company had to start training technicians all over again at Ford's British and German foundries.

Ford was concerned that Ferrovorm did not understand the problems it faced; it asked two engineers from Britain to travel to South Africa and conduct a detailed survey of the South African foundry industry. The head of manufacturing engineering and the senior process engineer stayed for five to six weeks to check out all aspects of their foundry and others; they decided that Ferrovorm could produce engine blocks, but recommended that the V-6 block be produced in green-sand.

In early 1974, Ford decided it needed an expert foundryman to help out, since it would soon need local supplies of more components in view of the higher local-content requirements. It requested assistance from the Latin American Group, and the vice president, in consultation with the technical director, decided on the manager of the foundry in Argentina, who was then in Rio to solve a problem there. After a resetting of priorities among the three companies, the foundry manager was detailed for a month to South Africa. He arrived in 48 hours and stayed one month, with South Africa arranging to get him back later for two months, hopefully, after he took care of an assignment in Mexico. His expertise is such that he can solve a specific problem in a matter of hours while it might take others days or weeks, if they could solve it at all. For example, Ferrovorm had experienced considerable problems with its resin sand system that led them to believe it would have to be thrown out; but the visitor said it was first-class and needed only to be used correctly; he had it worked out in a few hours and soon instructed the local engineers and operators. The solution to the problem helped other auto companies within a short time.

The main item causing problems was the inability to produce cores on the German machines; the visitor, using his Argentinian experience was able to solve this. He did so by adopting a simple procedure—opening holes in the core box to release air that was resisting the flow of sand as it was forced in. This procedure was known to others but rejected by Ferrovorm as unworkable; the visitor knew it would work because he had used it in prior situations. The result was an increase of productivity by 30 percent in the Ferrovorm foundry. Samples are now coming out, but Ford has still received no engine blocks from Ferrovorm, despite spending $1,000,000 on tooling to fit the machines bought by Ferrovorm.

While the visitor was at Ferrovorm, he was asked to visit another (private) foundry in South Africa to assess its capability of making castings. He got copies of the Chrysler foundry layout and that of Ford's Argentina foundry that would meet the local foundry's needs. After his visit to Mexico, this expert will also go to Australia, where Ford has bought out the supplier foundry, which will be run by one of the British engineers who had investigated the South African situation. These trips illustrate the wider help that Ford extends inside and outside its family.

One of the problems with Ferrovorm was its inability to accept technical assistance from others. This problem reflects what sometimes happens when a process is bought in a lump by a company that has little experience in an industry or does not have personnel who are interested in or capable of using it. This was partly a result of its constant turnover of personnel—at the end of two years, it had only one engineer who was in at the beginning—which has forced continual retraining programs.

Commonisation—Axles

In order to expand self-sufficiency by reducing costs of supplier companies, the government has encouraged commonisation of parts and components among auto companies. Ford found it could not commonize even among all of its own models, but some components can be commonized among models and across auto companies. In response to the government's policy, Leyland set up a diesel plant for commonized engines, but it could not get agreement among the companies, so it closed up and shipped the equipment out. Commonisation requires retooling, reengineering; and, if carried far, it cuts the industry off from the parent companies and overseas developments. But it can work with substantial technology exchanges.

Borg-Warner has succeeded in some axles because it knows how to accept and extend technical assistance. It is an international company and is used to working closely to others' engineering specifications. It was already in Australia producing for the Falcon and Fairlane. It came into South Africa and offered commonisation of axles, which would require an adaptation of the Cortina and Escort. Ford-SA was able to make the changes with Australian assistance on the power train design, from the Gleason company on gear cutters, from Borg-Warner in the United States, and from Ford-Britain on the design criteria.

The axles failed at first, and consultation was required with Dearborn, which lead to extensive tests and revisions. New results were found satisfactory and modifications were made to satisfy GM in South Africa, so they could join in axle commonisation.

To make certain that the final results suited Ford–SA, an engineer went to Australia to visit Borg-Warner there and examine each step in production. A Dearborn official had been to Germany to solve problems in axles previously and knew some of the problems Ford–SA would face; he advised a set of controls to improve the quality. Borg-Warner had set up to manufacture all but the gears, but quickly adopted the advice of the Dearborn and Ford–SA engineers, including advice on the deskilling of manufacturing processes so as to use blacks (an absolute necessity, given the scarcities in the labor force). Some of the changes, which were to meet requirements of the South African roads, result in a reduction of warranty problems to about one-eighth of previous experience in Australia for the same axles. Noise was also reduced. Success was achieved with continued, full, and cooperative exchange of information among Ford affiliates and between them, the supplier, and a competitor.

REPAIR AND MAINTENANCE ASSISTANCE

Service is one of the more serious problems in the auto industry in South Africa, for mechanics are scarce and the turnover in garage mechanics is high. The turnover requires frequent retraining on particular vehicles, and the high demand is opening opportunities for coloreds and blacks in garages. This has been offset somewhat, however, by the reduction in accidents due to the dropping of speed limits as a result of the oil shortage.

Ford has developed a "Registered Technicians Program" under which it sends out monthly booklets as materials for a correspondence course. The materials are based on an exchange of information with Dearborn and liaison with Britain on their cars, supplemented by visits from Ford–Europe officials. However, the packet is prepared and distributed at South Africa. It is followed up with on-the-spot visits to garages and dealers' service departments.

In addition, Ford offers one-week training courses in Pretoria and Port Elizabeth every two weeks that cover all aspects of mechanical and electrical maintenance, transmissions repair, and acquisition of parts and accessories (all but glass and radios). Between 12 and 20 persons attend each session.

In 1974 these courses covered topics such as diesel fuel injection systems, automatic transmission, powertrain for cars, noise-vibration-harshness, electrical repair, brakes and steering for cars, engines and tune up, water and dust leaks, general apprentice course, and new model service features.

For customers such as the governmental departments (police,

and so forth) and for truck and auto fleet owners, special courses are provided of a one-week duration on all aspects of maintenance and repair. These are provided on demand and usually relate to new models. However, some courses are also offered to engineer-trainers who must constantly instruct newly hired mechanics.

During 1973, over 1,000 mechanics attended all training programs conducted outside the company by Ford technicians; 1974 showed an increase in enrollments. Within this 1,000, blacks and coloreds were being trained in special skills, such as gear boxes and carburetors but not on the entire vehicle. These workers would have already been hired in a garage or in a service department of a dealer and would be seeking training for a specific task. In addition, the company was planning to put a travelling instructor on the road permanently to train nonwhites; over 100 are expected to be contacted on the first round trip. These trainees are likely to become a source of future employment in the factory, after experience in service departments around the country.

Owners themselves can buy a "workshop manual" from their dealers by which to repair their own vehicles; the company estimates that about 25 percent of all repairs are done by individual owners.

Productivity in repair shops is critically low, and Ford tries to improve it by sending them the instructions on workshop layout and work-flow used in Britain and the United States and by providing specifications on repair stations after diagnostic work. The breakdown of jobs into specific tasks opens opportunities for blacks as mechanics. These aids are especially useful to shops with more than 60 cars per day to repair; when followed, they have increased productivity as much as 20 percent.

The transfer of skills in repair and maintenance is illustrated by the fact that the automatic transmission was early treated as a "black box" by dealers; it was simply removed and sent back to the factory when there was any trouble with it; in the meantime a new or rebuilt transmission was put in the car from the dealer's parts inventory, but this was expensive since it required dealers to stock the transmissions. Now the company is training mechanics for such repair by running clinics that make use of video tapes and other instructional aids.

Another type of technical assistance to dealers is provided in the Best Ideas Distribution System (BIDS) by the Dealer Service Bureau (DSB). A computer service is provided to help dealers run their parts inventory control better and to keep track of their accounts receivables so as not to lose money. The dealer captures the parts sales data on punch cards, sends them by mail to the Ford computer center, and the results, which are flown back the next day, tell him what he has in inventory and help him estimate orders in advance as well as costs of maintaining his stocks. This system also permits the swapping of inventory among dealers when one develops a shortage, for the status is recorded at a single center. This service has cut inventory costs to dealers by 20 percent. It will be expanded to cover financial data (receivables and payables) and to workshop controls and payrolls, so as to improve dealers' profit picture.

This computer service is used also for demand estimates by each dealer. Each puts his demand forecasts into the program, and Ford then builds its own forecasts out of the combined information, which results in a better match of dealer forecasts and production of specific models by Ford-SA. A problem still remains because production lead time is five months and order lead time is only two months from dealers; therefore, Ford has to make its own longer-range demand estimates, for it can respond to shifts in dealer demand only "at the margin" within the limits of existing CKD units at the factory.

The BIDS program was developed for Brazil by Dearborn to avoid production in advance of orders by building only against orders; but it needed demand forecasts to do this. The program is more suitable to Brazil than to South Africa for the Brazilian lead time is less because there is less reliance on imported CKD units. South Africa redesigned the computer print out to meet its needs better; the system has smoothed its production and reduced its "inventory" production as compared "to order."

In addition to the above services to dealers, Ford has given assistance on warehousing, shop layout, and show room layout and facilities. It has developed a modular layout for warehousing that can be put through the computer to show the best one for each dealer; this idea is being examined by the manufacturing staff at Dearborn.

Opportunities for dealers are continually opening up, and with the potentially expanding demand by blacks, Ford can see a time in the near future in which black dealerships will be formed; GM already has one in Capetown. Eventually also, black enterprises will be developed as subsuppliers to supplier companies, since this is being encouraged by the government.

PRODUCT DEVELOPMENT

Ford-SA has changed models when Europe or Australia has: (British and Australian right-hand drive matches that in South Africa). The next models will include modifications of the Escort to achieve still greater integration among European affiliates.

Technical assistance has been a continuing flow—not only on existing facturing processes, to suppliers, and dealers but also on new models, as they are developed by the parent company or within another affiliate. In South Africa, this aspect of technology transfer is both accelerated and altered by the impending increase of local content. What products can be adopted will depend largely on the ability to obtain local supplies of components. which can slow the adoption of new models; and the necessity to rely on these supplies will accelerate acquisition of technology on production of components.

The rise in local content projected for the next phase—after 1978—is likely to have effects similar to those in other countries with high local content: a retention of models that are discontinued by the parent or other affiliates; development of unique South African models, such as the Cortina Big Six, which is not produced abroad; and an increased commonisation of components among suppliers within South Africa to reduce costs. Commonisation will require new technologies and designs and exchanges of technology and engineering data among major companies.

It can readily be projected that the number of models each company offers will be reduced and the twelve companies with sales of less than 10,000 units per year will be pushed hard to stay in the market at all. The necessity to rely on higher-cost local suppliers will force losses on many, who may have to be subsidized by the parent company or pushed out of the market. But, improvement in costs with larger volumes of component supply is not likely, for suppliers are limited by their own capabilities; for example, there are only two companies making exhaust systems for the entire industry, and their ability and capacity is limited which forces compromises in quality, volume of production, and supply dates. The supplier of radiators still has problems with leakages because of poor inspection and quality control; to go to still higher volumes in mass production will require substantial technical assistance in quality control. Similar, but more difficult, problems will arise with items that are now produced abroad (and imported) because of their complexity. The costs of technical assistance will rise substantially, and other companies will have to ride on that provided by Ford and GM if they are to survive.

Ford-SA has built up a substantial capability in product development and modification through attracting European engineers and technicians; 60 to 70 percent of the personnel have immigrated from

Ford-Europe affiliates; only one engineer has been seconded from the United States or Europe at any given time. This capability is supplemented by being tied into computer information on technical developments and market trends in Britain through Ford-SA's liaison engineer in London. Any new development or modification must be checked out with Dearborn to see if it is "good for Ford" as well as "good for Ford-SA."

The fact that South African demand (and the peculiarities of the tax system) brought relatively high sales of pick-up trucks lead to the development of a Cortina pick-up not made anywhere else by Ford. Ford-SA had moved out of the Zephyr-Zodiac into the Taunus and Granada with a 6-cylinder engine. It took the 4-cylinder engine out of the British Cortina and put in the Granada's V-6. (Australia had made a similar change, using the "in-line" 6; each was already producing different 6's and simply used the engine it had. The two had exchanged information on modifications necessary to support the heavier engine and on adaptations when Germany made changes in the engines.) Two changes were made to strengthen the suspension and to improve the cooling system of the engine.

The Cortina pick-up was made from the model with the larger engine to meet competition. The modification was to cut the body of a 2-door sedan in half behind the driver, design a panel for the back of the cab, change the chassis by cutting it in half and splicing a pick-up type chassis onto the front frame, strengthening the front axle, making a heavier rear axle and drive shaft, and designing a pick-up body and tailgate. The major problem was the stress-strength of the spliced chassis. A German engineer was flown down for testing, and a Ford-SA engineer was sent to the United States for final evaluation and testing prior to approval for export. (Export would mean that Ford Motor Company stood behind the vehicle, whereas sales only in South Africa would imply only that Ford-SA was responsible.)

It does not appear that there is any significant market for the Cortina pick-up or the Cortina-6 outside of South Africa, for no other affiliate has shown any interest. Australia has its own pick-up out of the Ranchero, and the Philippines and Taiwan are producing a much lighter vehicle—the Fiera. In addition, export is unlikely because of the absence of high-volume production of the model, lack of export incentives by the government, and lack of regular shipping schedules—all of which are needed for export.

The next modifications of the Escort, produced in Britain and Germany on an integrated basis, will be offered to Ford-SA, which then benefits from the design and engineering resources devoted to the new model. If the South African company were autonomous and self-contained, it would not be able to maintain either a modern model line or to introduce major technical improvements. There is a cost for this degree of integration, however, in that production is subjected to possible interruptions in supply of components from either Britain or Germany. There are other problems during the transition, for the British Escort was changed slowly in anticipation of the new model, thereby giving South African suppliers a fit in making their components, which did not interface correctly with new imported parts. South Africa had to persuade Britain to continue the old components until the total shift could be made, at least for supplies to South Africa. It was supported in this request by Australia, which faced the same problem. The Cortina suspension was also modified ahead of the introduction of the new model, as with the Escort, and both countries had to request a delay in order to adjust.

By mid-1976, a new model for the Cortina will be coming out, and Ford–SA will face the necessity of new investment and new technical assistance. Technical assistance starts with the building of a pilot model; three manufacturing engineers have already visited Ford–SA to see what problems will arise in the change over. This technical assistance must be passed on to suppliers at the early stage of die-model making. This is a highly skilled trade, requiring an exceptional degree of dexterity, the ability to read complex drawings, and translate the designs into clay and ultimately fiberglass, wood, and finally metal. These are new skills in South Africa, the costs of which are prohibitive (including materials and equipment) for any small supplier making stampings. Ford assists on all aspects.

Minor changes in the product have occurred as a result of engineering developments in South Africa as well as in Europe. For example, changes in the fuel-tank capacity, suspension and springs, suppression of the sparking system in the engine to prevent interference with TV, safety and pollution equipment, and changes in tire standards from 6-14 to 6-90 to meet requirements of the South African Association. In another instance, Ford–SA developed an air conditioning system with a local supplier to meet a problem of hot air being drawn over the engine into panel vents and thwarting the air conditioning; this effort also involved a two-week visit by a German engineer.

All such developments are reported to the Latin American Group and through the Worldwide Engineering and Research group in Dearborn. As a result, some European companies have requested data on the South African solution to a problem with the Granada drive-line.

DISSEMINATION

Apart from those directly connected to the motor industry, technology is disseminated merely through use of acquired skills in everyday life or in changes in lifestyles of employees. In addition, shifts in personnel out of Ford into other companies disseminate technology throughout the motor industry.

The highest level transfer into other companies occurred a few years ago when the manufacturing director was bid away by a competitor after his being with Ford–SA for 41 years. He carried with him considerable technical skills and managerial know-how in using Ford methods and systems. Both Toyota and Datsun (locally owned licensees of the Japanese companies) have pirated Ford officials, partly because they find it so difficult to obtain technical assistance and managerial know-how from the Japanese licensors. The pirating has extended through the plant manager level, to quality-control technicians, and into the work force. In these two companies it was rumored that "every job worth holding" was occupied by a former Ford man.

Companies coming into South Africa during the past five or six years have also sought out Ford (and GM) managers and given them attractive promotions and 50 percent increases in salaries, which is hardly adequate compensation for the expertise bought and certainly no compensation to Ford. VW also obtained a number of black hourly workers in 1972 from other companies as a result of new legislation permitting greater mobility of workers. The complete record of such transfers of salaried employees is shown for 1971–74 in Table 3-4.

Such transfers are less frequent to supplier companies, for the manufacturing companies pay higher wages. However, an engineer in Ford-SA's product development (grade 9) became a general manager of a supplier company, though he was poorly prepared, simply because he had fifteen years of experience with Ford. About half of the turnover of mechanics is to small repair shops, which are in desperate need of such skills. Also some go into small business requiring particular skills; for example, one electrical technician was paid 30 percent more than he was receiving at Ford to work for a small electrical concern.

On a different level, technology is transferred into everyday life through several of the workers becoming "backyard mechanics"; this helps explain the high percentage of repairs reportedly done "by owners." Several workers also reported changes in their aspirations as a result of working for Ford. One Asian worker has been

Table 3-4.   Ford-South Africa Employees Who Transferred to Supplier Companies or Competitive Automotive Companies, 1971-1974

Position in Ford South Africa	New Employer	Position with New Employer
Superintendent Production	Grosvenor Motors	Service Manager
Truck Marketing Manager	Leyland	Truck Marketing Manager
District Manager	Peugeot	Zone Manager
District Manater	Steyns Ford	Marketing Manager
Cost & Inventory Analyst	Roslyn Motors	Financial Analyst
General Foreman—Production	Datsun	Superintendent Production
Truck Representative	Toyota	Region Manager
Tractor Field Specialist	Ford Dealer	Vehicle Salesman
Buyer—A	Anikem	Buyer
Systems & Methods Analyst	Datsun	Systems & Methods Analyst
Sales Training Manager	Leyland	Sales Training Manager
Truck Sales Analyst	Leyland	Market Analyst
Superintendent Production	Car Distributors & Assemblers	Production Manager
Service Zone Manager	Eriksen Ford	Service Manager
Marketing Strategy Analyst	Grosvenor Ford	
District Manager	Van Zijl & Robinson	Accounts Executive
Plant Vehicle Scheduling Clerk	Datsun	
Performance Parts Co-Ordinator	Volkswagen	
Process Engineer	Car Distributors & Assemblers	
Parts Control Analyst	Chrysler	
Report Analyst	Volkswagen	
Foreman—Production	Illings Earthmoving Equipment	
Laboratory Engineer	Dulux	
Computer Control Analyst	Motor Assemblies	
Dealer Auditor	General Motors	
Parts Inventory Analyst	Mazda	
Industrial Engineer	General Motors	
Foreman—Maintenance	Volkswagen	
Cost Analyst	Shatterprufe	
Foreman—Production	Shatterprufe	
Purchase Analyst	Datsun	

Table 3-4. continued

Position in Ford South Africa	New Employer	Position with New Employer
Foreman—Machining	Citroen	
Product Engineering Designer	General Motors	
Accounting Assistant—B	Guestro Industries	
Security Officer	General Motors	
Foreman—Production	Castor Love	
Product Test Engineer	General Motors	
Process Engineer	General Motors	
Unit Supervisor MPPO Stores	Guestro Industries	
Industrial Sales Engineer	Grosvenor Motors	
Maintenance Foreman	Datsun	
Laboratory Engineer	General Chemicals	
Process Engineer	Repco-Wispeco	
Industrial Engineer	Leyland	
Supervisor Local Parts Control	Citroen	
Option Merchandising Specialist	Volkswagen	
Supervisor Specification Cataloguing	Illings	Superintendent Production
Superintendent Production	Leyland	Zone Manager
Industrial Sales Manager	Leyland	Managing Director
General Manufacturing Manager	Motor Assemblies	Buyer
Purchase Price Estimator	Besta Engineering	Service Zone Manager
Service Operations Manager	Illings	Material Planning Manager
Project Leader	Motor Assemblies	Personnel Manager
Manager Personnel Services	Motor Assemblies	Product Quality Manager
S. Q. A. Manager	Maritime Motors	Vehicle Salesman
Programming Co-Ordinator	Motor Assemblies	Systems Analyst
Systems Data Analyst	Thompson-Romco	Buyer
Buyer—A	Motor Assemblies	Personnel Supervisor
Supervisor Plant Protection	Motor Assemblies	Design Engineer
Product Design Engineer	Illings	Resident Engineer
Principal Design Engineer	Motor Assemblies	Production Manager
Production Manager	Atkinson-Oates	Vehicle Salesman
Zone Manager Tractor Sales		

Table 3-4.   continued

Position in Ford South Africa	New Employer	Position with New Employer
Manufacturing Operations Manager	Motor Assemblies	Manufacturing Operations Manager
Production Control Clerk	Illings	
Industrial Engineer	Ford Dealer	
Procedures Analyst	Guestro Industries	
Foreman—Material Handling	Datsun	
P & A Records Clerk	Car Distributors & Assemblers	
Co-op Trainee	Shatterprufe	
Cost Analyst	Datsun	
Layout & Guage Technician	Volkswagen	
Purchase Price Estimator	Besta Engineering	
Cost Analyst	Firestone	
Industrial Engineer	Leyland	
Design Draughtsman	Chrysler	
Foreman - Maintenance	General Motors	
Computer Programmer	Motor Assemblies	
Statistical Quality Control Analyst	Chrysler	
Computer Programmer	Motor Assemblies	Computer Programmer
Vehicle Engineer	Lawson Motors	Product Design Engineer
Senior Buyer	National Die Cast	Purchasing Manager
Purchase Price Estimator	General Motors	Purchase Price Estimator
General Foreman Production	Leyland	General Foreman Production
Marketing Research Manager	de Villiers & Schönfeldt	Market Research Manager
Supplier Quality Assurance Representative	Motor Assemblies	Supplier Quality Assurance Representative
Corporate Project Manager	Leyland	Finance Manager
Supervisor Indirect Laboratory Methods & Material	Leyland	Supervisor Industrial Engineering
Foundry Liaison Engineer	Ferrovorm	
Field Sales Manager - Tractors	Leyland	Field Operations Manager
Supervisor Supplier Quality Assurance	Leyland	Supplier Quality Assurance Manager

Table 3-4. continued

Position in Ford South Africa	New Employer	Position with New Employer
Unit Supervisor Computer & Inventory Analysis	Motor Assemblies	Supervisor Product Control
General Foreman Production	Motor Assemblies	Production Superintendent
Foreman Material Handling	Citroen	Supervisor
Dealer Systems Analyst	Toyota	Systems Analyst
Tractor Dealer Planning Co-ordinator	Leyland	Zone Manager
Buyer—B	Leyland	Buyer
Cost Analyst—Senior	Leyland	Cost Analyst
Unit Supervisor Scheduling & Programming	Motor Assemblies	Programme Manager
Project Engineer	Volkswagen	Project Engineer
Buyer—A	Citroen	Senior Buyer
Supervisor Technical Training	Pascor	General Manager
District Manager	Leyland Dealer	Sales Manager
Supervisor Quality Assurance Representative	Leyland	Supplier Quality Assurance Supervisor
Process Engineer	Motor Assemblies	Process Engineer
Product Design Engineer	Illings	Product Design Engineer
Product Design Engineer	Gabriel Mufflers	Product Design Engineer
Supervisor Tool & Field Services	Leyland	Field Service Manager
Special Markets Sales Manager	Malbak	Region Manager
General Superintendent Production	Motor Assemblies	Production Manager
Industrial Sales Engineer	Perkins Engines	Sales Engineer
Sales Training Co-Ordinator	Mandy's Motors	Salesman
Superintendent Material Handling	Leyland	Finance Manager
Tractors Zone Manager	Leyland	Zone Manager
Purchase Price Estimator	Leyland	Buyer
General Tractor Manager	Eastvaal Ford	Marketing Manager
Supervisor, Market Analysis	Toyota	

Table 3-4. continued

Position in Ford South Africa	New Employer	Position with New Employer
Computer Programmer	Motor Assemblies	
Sales Analysis Co-Ordinator	Leyland	
Industrial Engineer	Dunlop	
Computer Operator	Volkswagen	
Product Quality Assurance Engineer	B.M.W.	
Buyer—C	Citroen	
Production Control Co-Ordinator	Repco	
Industrial Engineer	Leyland	
Foreman—Material Handling	Citroen	
Layout & Material Handling Engineer	Chrysler	
Industrial Engineer	Leyland	
Buyer—B	Leyland	
Statistical Quality Control Analyst	Volkswagen	
Cost Analyst	Citroen	
Buyer—B	Leyland	
Sales Analysis Clerk	General Motors	
Foreman—Inspection	Girling	
Industrial Engineer	Girling	
Supervisor Vehicle Design	Pascor	
Truck Marketing Strategy Analyst	Roderick & Brook	General Manager
Fleet & Leasing Manager	Leyland	Service Manager
Purchase Price Estimator	Leyland	Field Operations Manager
Pre-Production Control Manager	Motor Assemblies	Purchase Price Estimator
Business Management Consultant	Toyota	Product Control Manager
		Business Management Consultant

Table 3–4. continued

Position in Ford South Africa	New Employer	Position with New Employer
Buyer—A	Messina Group	Buyer
Design Engineer	Illings	Design Engineer
Unit Supervisor Direct Labor & Methods	Leyland	Supervisor Industrial Engineering
Product Quality Assurance Engineer	Rosslyn Motors	Quality Control Manager
Truck Marketing Strategy Analyst	Aimco	Salesman
Tractor Dealer Planning Co-Ordinator	Imco Agricultural Implements	Assistant Sales Manager
Tool & Field Representative	Motor Assemblies	Supplier Quality Assurance Representative
Assistant Controller	Carborundum	Finance Manager
Product Design Engineer	Volvo	Product Design Engineer
Heavy Truck Marketing Manager	Leyland Dealership	Truck Sales Manager
Parts Release Field Representative	Leyland	Parts Release Representative
Cost Analyst	General Motors	
Foreman—Salvage	General Motors	Supervisor
Buyer—B	Messina Group	Buyer
Material Handling Engineer	Leyland	
Marketing Administration Co-Ordinator	Lippstreu Ford	
Process Engineer	General Motors	
Buyer—B	Motor Assemblies	
Buyer—C	Volkswagen	Buyer
Customs Entry Specialist	Balco	P & A Representative
Buyer—C	Motor Assemblies	
Procurement Timing Analyst	Chrysler	
Parts Zone Manager		

with Ford only five months, having come from a repair shop where he lost his job; he is seeking to up-grade his skills by getting some training in transmissions and differentials so that he can emigrate to the United States as a Ford worker; in the meantime he maintains his own car, his parents' and some neighbors' by making minor repairs.

An Asian section leader in the flatting area entered the company some years ago as a flatter; he became a spray painter after six months of training and spent fifteen years in that position; after a year of leadership training he was made a section leader. He has found this training exceptionally useful in assuming leadership roles in his church, teaching the youth of the church, and in his community organization, where he has taken an active role and helped manage its activities. Previously he knew nothing of how to organize others. He would like to become a mechanic, but his lack of education prevents it; although Ford would pay for his training, he has too many community commitments to take the time at present. He has, however, trained spray painters and has two group leaders under him who train new flatters. He has a pride in his accomplishments and would like to reach higher.

A black shop steward in the engine plant is also a member of the Liason Committee from his section. (Dissatisfaction that might arise from the blacks is mitigated by the Liaison Committee, which is required by the Bantu Labour Relations Act and in which each section is represented and all grievances are aired. The committee meets monthly to respond to suggestions or requests by the Bantus and to permit management to explain its objectives and methods.) He is an inspector on oil pressure and gaskets in the engine. He has had ten years with the company—first on machining of the engine, then on assembly, back to machining, and then to inspection. He is learning to be a foreman, but needs a Standard 10 education; he is now completing Standard 8 and will seek to graduate in a couple of years. He faces the same dilemma that many others do of deciding whether to seek more education on his "off hours" or to accept overtime work with its higher pay. (When overtime is required because of production scheduling, it tends to interfere with studies.) Most workers choose the money now rather than education for higher responsibility and pay later. In addition, he reported, almost all mechanics moonlight on repair jobs with small garages, as do the spray painters, who are in great demand in body shops. In his group, he stated, there is little turnover, for work opportunities are the best in the auto industry in South Africa, and many who leave for other companies seek to return in a few months.

Chapter 4

Ford-Lio Ho (Taiwan)

Ford-Lio Ho is a joint venture between Ford Motor Company and the former Lio Ho Automobile Industrial Company formed by the Tsung brothers, who made their fortune in the textile business in Shanghai and Taiwan. Between 1968 and 1972 Lio Ho was a licensee of Toyota; it sold a 70-percent interest to Ford in 1972, with the consent of the Taiwanese government. Lio Ho was one of six local companies, each of which had a licensing agreement with a Japanese auto maker. At the time it ceased production of Toyotas, it accounted for about 13 percent of the auto market in Taiwan.

Toyota had begun making overtures to mainland China in 1971, and it appeared that it might let its license with Lio Ho lapse. Henry Ford had become interested in a venture in Taiwan during a visit in 1971. Ford took the initiative to fill in a potential gap, and Toyota did nothing to discourage the negotiations. Only after Lio Ho was dropped did Toyota discover that mainland China was not as attractive as it appeared; it later sought another affiliate on Taiwan.

Since Lio Ho had four years of production experience with Toyota prior to becoming an affiliate of Ford, this study provides a contrast between technical assistance provided under a license to an independent company and that by a majority owner. The *Taiwan Trade Monthly* (November 1973), in commenting on prospects for the auto industry, raised questions about the benefits of independent licensing, under which all six companies had operated. It observed that Taiwan-made cars cost much more than foreign counterparts, that quality was below standard because of parts made locally with tools, instruments and specifications not up to standard, and by workers with too little technical training. It asserted that "Foreign investors may help raise the standard. Ford technicians have worked with Ford Lio Ho subcontractors to help them improve their products and institute quality control procedures."

INITIATION OF PROJECT

Lio Ho was a new company, formed specifically to produce Toyotas under a license agreement. The plant was constructed with this sole objective; workers, who were mostly drawn from the engineering and technical school run by the Tsung brothers for their textile and other interests, were sent to Japan for training. Seven engineers were sent to Japan to learn their jobs; they were taught some foundry skills and assembly of the car and engine (Mini Ace and Corona—the agreement was limited to these two models and Lio Ho could not add others on its own and later felt Toyota would not agree to any change). None were taught inspection or analysis of problems. They were taught techniques, such as spray painting and machining of engine parts, but only the *one* operation needed to do the job.

Having learned the production processes, they returned to Taiwan, and one Japanese supervisor was named liaison in Japan. Japanese engineers were sent to help set up the engine line and three Japanese engineers were sent to stay in Taiwan: a chief engineer (who was a retired Toyota manager), one in the engine plant, and one in the foundry; in addition, three technicians remained to check on problems of metal, paint, and inspection in the assembly operation. These men solved any problems by themselves or obtained assistance from the sole distributor in Taiwan associated with Toyota. (All sales by Lio Ho were to the sole distributor, which had its own retail outlets and repair and maintenance service; Lio Ho was, in effect doing contract manufacturing only.) There was little communication with the company in Japan, since Telexes were considered too expensive by the Japanese personnel; Lio Ho had to "pull out" of the Japanese whatever it needed.

As production began, parts validation was carried out in Japan save for a small number that were inspected in Taiwan. Two complete engines were sent back to Japan for testing after assembly was initiated. Corrections were sent back on each and every part. When the Corona RT-80 was introduced (after a year and a half of production), Toyota sent three men to train the Chinese on inspection of the assembly operation; they stayed two or three months. No more than five such officials visited Lio Ho in its four years with Toyota.

Two significant gaps in training were noted by the Lio Ho engineers, besides that of problem analysis. No attention was given to manufacturing engineering and none to supply management; these omissions resulted from the fact that Toyota made all these decisions and left Lio Ho simply to carry out operations on whatever models were sent CKD (the model range remained simply the Mini-Ace and three of the Corona series).

Lio Ho felt greatly restricted in its activities; its operations were, in fact, closely circumscribed. Its personnel learned only the assembly of the CKD units. It would have needed, by the testimony of the engineers, much more information to be able to handle the problems raised or to get into diverse models. The engineers would have liked much more technical assistance, and the managers wanted to go into a higher-class Toyota, but they were prevented on both scores. They were, in their view, only in "kindergarten," learning the first steps; their only consolation was that their prime competitor (Yue Loong) had been treated the same way by Nissan since 1953, though it did have independent distribution facilities.

The benefit to Toyota lay not in the royalty fee paid by Lio Ho—which was not large—but in the price of the CKD units and the fact that by channeling sales through its exclusive distributor it got greater market penetration for *all* Toyota models—those locally assembled and imported. Lio Ho concluded that it could not obtain from *any* auto company what it would need to become an effective producer of the line of cars (making and selling on its own models), so long as it remained a licensee. It was, therefore, ready to negotiate when Ford came along. With Ford it has doubled its market share to 30 percent, while Yue Loong with a 15-year headstart, retains 50 percent. Also, the arrangement with Ford set a new pattern of multiple independent dealers, thereby rejecting the exclusive distributorships normal in Taiwan. Before Ford entered the picture, however, it made a thorough study of the auto industry for the Taiwanese (ROC) government, which helped shape its own decisions.

Ford Study

The ROC government had been interested in getting foreign investors in the auto industry rather than mere licensing but did not want to give up all ownership to foreigners. It wanted Chinese to keep at least a minority position. In order to frame its own policies, and incidentally induce serious investigation by a foreign company, the government asked Ford to study the future of the auto industry and make recommendations on the best policies. This became a dual "learning and technical assistance" project for Ford, for it was done in cooperation with officials of the Ministry of Economic Affairs. Entitled "The Auto Industry in Taiwan" and dated June 1971, it focused on the problems of (a) local content requirements, (b) the penalty costs from local manufacture, (c) impact of taxes on costs, (d) the deficiencies of local manufacturers, and (e) the vehicle market. These led to an assessment of future alternatives for the industry, backed by an extensive review of the availabilities of local supplies of parts and components. The major conclusion was that the industry should be opened to foreign investment because foreign licensing had not lowered costs nor supplied potential market demand.

Ford's assessment was that the way in which the licensed production was working actually raised the costs of the cars above what they should have been. When cars were imported completely assembled, the Japanese companies sold them at (lower) export prices based on variable cost, to which duty was added. But when local assembly was instituted with CKD and some local parts, the deleted parts were priced at variable costs and CKD parts at *full* cost, which actually raised the cost of the imported car significantly. In addition, a high effective duty of over 50 percent, which was put on the CKD parts to induce local manufacture, raised the costs of local assembly even higher. In this system, the Japanese CKD exporter sought to recoup his entire former profits from the CKD parts shipped and (in the case of Lio Ho) received all the sales profits—since it had no equity interest in the licensee.

The report showed, also, that significant deficiencies existed among auto companies in Taiwan in the technical aspects of production, in marketing, finance, product planning, and general management. It proposed that, if licensing was to continue, specific agreements be concluded on precise technical assistance to be extended by the Japanese companies—this, however, was not done.

With the decision of the government to open the industry to foreign investment, Ford proceeded with its own feasibility study and negotiation with Lio Ho. The present president—then assistant general manager of Philco-Ford in Taiwan—was given the responsibility for the negotiations. Since this was a negotiation of a joint venture with a partner composed of older Lio Ho officials having little knowledge of English and some younger officials with English knowledge, the presence of the Philco-Ford official, who was a Chinese-American was a critical factor in bringing the project to a successful conclusion. He was able to help resolve differences through his greater appreciation of the situation as seen by the local nationals, without which the deal would probably not have been consummated and the technology would not have been transferred. Once negotiated, the project was submitted to the central staffs at Dearborn by the officials at Ford-Asia-Pacific; few changes were made at Dearborn because FASPAC had already gone through several other projects in the area and knew what to do.

Expansion of Ford-Lio Ho

A basic decision was made at FASPAC to more than double the capacity of the plant, up to 15,000 units per year, eventually. The specified plan indicated the domestic market demand and the extent of "complementation"—that is, export of components to other Ford affiliates in the region. It was contemplated that Ford-Lio Ho would be the supplier of engines, which in turn would give it a credit under the local-content requirements to the extent of these exports. Production was projected at 24 Cortinas or Escorts per day; 20 Fieras per day, and 86.5 engines. (This compared to 12 Coronas and 8 Mini-Aces previously.)

During May thru December 1972, program groups in FASPAC specified the requirements to meet this plan. They made a series of visits to Chungli (the site of the plant some 20 miles from Taipei) during June and November 1972. These visits included personnel from product engineering, manufacturing engineering, accountant, supply, finance, and marketing. Expansion of the plant was begun in January 1973, and machinery equipment and tooling was ordered through August. The building expansion was completed in March 1973; the first Cortina came off the line in March 1973, followed by Escorts in July, and Fieras in November. The first 30 Cortinas and 10 Escorts, and 20 Fieras were 100 percent imported CKD (except for tires, batteries, and paint), but local suppliers came in promptly thereafter with springs, seats, carpets, and so forth.

In mid-1974 engines were still being assembled from completely CKD units, but the foundry had been producing prototypes for months and was trying to get up to quality for various parts of the engine. June 1972 was the target set for use of local castings from the foundry. (One of the reasons for the rushed schedule was the fact that foreign service personnel required by Ford-Lio Ho were charged at a cost of $130 per day—salary and expenses—and these were mounting up.) This was a tight timetable, but it was originally thought realistic because the work force was basically trained and basic equipment and layout were in place. In fact, more training and adjustment was needed than anticipated and the energy crisis delayed delivery of some imported equipment; however, by September 1974, local production of engines was launched.

The target of raising local content to 60 percent, from the original 40 percent, through credit for engine exports to Singapore, Thailand, New Zealand, and the Philippines was well on the way to being achieved by year end.

Sales took off quickly and surpassed the ability to deliver. By the end of 1973 (with less than eight months of operation) over 3,500 cars had been produced (compared with Lio Ho's 4,400 for all of 1972), and this figure was to be more than doubled in 1974. By mid-1974 the work force was over 600 men, many of whom were in training on the floor in preparation for full-scale operations. The Cortina sells at about $4,200, the Escort at $3,800, and the Fiera at $2,300. Even at these prices, the low volumes will probably not make Ford-Lio Ho profitable from merely domestic sales; its basic justification is as a part of the "complementation complex" for Southeast Asia, with each affiliate supplying others key components for similar vehicles, especially for the Fiera.

TECHNOLOGY TRANSFERS: CONSTRUCTION AND MANUFACTURING

Although the transfer of technology began in the prelaunch and even prenegotiation stages, the major impacts occurred in the modification of the plant and in the manufacturing techniques acquired by Taiwanese engineers and workers. The story is recorded in the sizable expansion of facilities, the changes in the foundry, up-grading of the engine plant, new approaches to assembly, and the introduction of the Fiera in cooperation with other Ford affiliates in Asia.

Construction Modification
In making the modifications of the Lio Ho plant to accommodate the expanded production lines and multiple models, FASPAC was able to use the Ford construction standards to a large extent, but not in all respects, because the existing building and facilities had to be used to the maximum to cut costs and to begin production as soon as possible. The paint shop was the one area that was least changed, for it was similar to the Ford standards, but it was still checked by FASPAC against the Ford model; in doing so, it found one process that was better than Ford practise and adopted it—a reverse technology flow!

Due to the relative lack of complexity of the Lio Ho operation, it did not include a plant and manufacturing engineering operation such as used by Ford. An engineering staff of three took care of minor contracting and liaison with government authorities, while processing and other manufacturing-engineering functions were the responsibility of foremen. The preprogram assumption by Ford that a substantial engineering base existed was not borne out. A foreign plant engineer who was assigned for only six months to oversee the expansion had to be extended to seventeen months in order to continue to provide the support needed; he even assumed a line position when one of the Chinese was ill. The experience of this foreign service officer indicates the expertise on which an affiliate can draw. He was assigned to the Ford–Australia team working on the Ford–Lio Ho program and was closely familiar with the objectives in Taiwan. He was an electrical engineer by background and had worked in the plant expansion in Australia with responsibility for a variety of facilities. Because of the urgent need for guidance at Chungli, he was given eight days' notice and sent "on his own." He found the Chinese to be very hard workers but that he needed to match their patience in order to succeed. They had operated a different system of contracting, with less emphasis on specifications, competitive bidding, and financial controls, and they were not goal-oriented, but tended to "fire-fight" problems rather than to schedule work so as to minimize problems later on. So, he had to give precise instructions in trying to help them understand the objectives and relate their technical background to practice on the floor.

The Taiwanese did not understand Ford procedures or the type of facilities needed to service the plant, and the transition was long and laborious. Having had some experience, they often considered that they knew the answers to a problem and wanted to solve it their own way. This made it necessary to permit them, for example, to do their own drawings for the expanded facilities and then to compromise between their drawings and Ford standards; it would not have worked merely to insist on Ford standards, for the Chinese would not have understood why and would have felt belittled.

However, in time, they have accepted the changed procedures and are more ready to adopt any new ones required. Their considered support for these systems has produced an irreversible change in their way of operation. For example, they participated keenly in the preparation by a local consultant of a critical-path network of all significant actions in the program and have used this tool to considerable advantage.

The early lack of foresight and planning on the part of the Chinese made it difficult to use even existing equipment. Ford thought it could keep many of the machines from Toyota days and rework them to start new processes. But there had been little provision for spares or stores for the machinery, which made such use problematical. Ford had to institute a preventive-maintenance system and a stores system for automatic ordering of spares for the machines.

What Ford was transferring to the Chinese was as much managerial as technical know-how and how to *mix* the two—that is, how to *manage* a technical operation. The Chinese had been given little managerial responsibility under the Toyota license and had never worked to a budget. It was, therefore, also critical to train the engineers in working to program and financial targets. At this early stage, the Chinese did not understand the total project scope or program objectives and therefore had difficulty in understanding the dynamic balance necessary between relatively autonomons departments and different affiliates that were cooperating. At times, the necessary "push" and "counterpush" was mistaken by the Taiwanese as fundamental hostility from the foreign service personnel, who were needed as an interface with the rest of the system.

This techno-management training was critical, however, for plant managers would have to work within money constraints on equipment to be purchased and repaired, and they had to know this before the foreign service personnel left. In one case, the Chinese received a lesson in sharp cost-cutting: a CKD fuel tank facility, which was to go in, was budgeted at $10,000 but the cost estimate they came up with was $20,000. The Chinese were induced to try to put it in at $10,000, which they did by using in-plant rather than outside labor and by adapting reusable materials and machines to put the facility in. (Now the Chinese are worried that, since they succeeded, the management will expect similar "miracles" later.)

A program orientation also requires a "follow-up" procedure, which the Taiwanese dislike. For example, if an engineer gave a technician an instruction, he was expected to complete it, and the engineer did not check back; however, if a problem developed, the technician would not report it, for it is not considered "courteous" to bear bad news to your superior. Therefore, the foreign service officer had to check the entire site every day, just to uncover problems. Similarly, the concept of preventive maintenance is foreign to them. But the Chinese plant engineer is beginning to get the idea and had even started shouting at the technicians in order to make certain that they carry out his orders and report back on problems.

During this expansion process, it was not possible to undertake any formal training of the engineers on maintenance or management and budgeting; the time schedule was too compressed to permit that luxury; teaching and learning had to be accomplished on the job. Ford found a number of engineers with varied and high-level capabilities, but they did have to be uncovered. For example, it found that the suspension weld guns had to be changed from the Toyota system. Every basic aspect of cable length, pivot, placement, and so forth had to be redesigned. Though it was difficult to get the Chinese to see the relationship of all aspects of the gun's use, once they did, they proceeded under guidance to redesign it with alacrity, and once they had done one, they easily finished the rest of them. The talents were there; they just had to be uncovered and used.

Assistance in the expansion went to minute, but necessary things, such as the absence of windows in the office building. Several alternative designs for the building's interior and exterior were proposed, including different materials for the interior. In the plant, the foreign service officer had to help buy the tools for production. The expansion could not have succeeded without a variety types of technical assistance, from minutiae to program planning and budgeting.

Foundry

Although a foundry was in place, its capacity had to be increased, an electric furnace had to be added, the sand processing had to be changed, and the product designs changed. The new foundry was designed to cast the engine block on the 1,100, 1,300 and 1,600 Kent engine series, the flange muffler, bearing caps, manifold exhaust and intake, water outlet connection, cover for the cylinder front, retainer for the rear oil seal, and the water-pump housing and impeller. By mid-1974, some of these were in production, while others were being cast for training purposes, and engine machining was being done to gain expertise. Full production began in September.

To prepare for the expansion of the foundry, a British engineer in Australia was detailed from February to August 1973—subsequent to a two-week crash operation in Britain to do the project plan for the foundry. The experience of this specialist illustrates the expertise made available to the Chungli facility—an expertise recognized by his being sent in November 1970 to South Africa to help with Ferrovorm. He had 14 years with the Ford foundry at Dagenham, two years of which he was as the manager of manufacturing engineering; he was then shifted to the same position at the Thames foundry and then made supervisor of process engineering at the Thames foundry after spending seven weeks in the United States to get tooling for the Thames operation; he had also done a complete project study on the Korean foundry after spending two months in Australia, three weeks in Korea, one week in Japan, and four days in Chungli to compare processes at low levels of production so as to scale down the Korean operation. In February 1973 he was seconded to FASPAC, where he spent 85 percent of his time on the Taiwan program and the rest on problems in New Zealand and Australia (where he advised on foundry planning for the future and on cost procedures).

During his 80 weeks of assistance to Taiwan, he spent about 30 in Chungli and the rest in Australia designing and purchasing tooling, designing the new furnace, hoists, conveyers, and so forth, getting quotations on equipment in Britain, Japan, and Australia (especially the tailor-made equipment). He was responsible for any changes in the basic design, for design of tooling and for ordering new equipment. He would give specifications to the tool manufacturer who had to fit these into existing Japanese machines. He had to lay out the proposed equipment along with existing facilities and processes. In one case, in setting up an alternative method of making cores for the castings, the tool manufacturer argued against the British specialists' proposal to vent the core boxes to permit more air to escape because it would be difficult to do and was both unusual to Ford practice and considered special to this particular foundry. But the specialist prevailed, and the method succeeded in making better castings.

More than one man was required to assist in the foundry, for it was being expanded from production of two types of engines and 12 sets of castings per day to 86.5 sets per day on three types of engines. To hasten the shift, the expansion was designed so as to keep all but 5 percent of the prior facilities. The shifts were raised from one to two and the equipment in production more than doubled, thereby raising capacity four to six times.

To accomplish this task, Australians were needed on the initial planning stages in Australia in March; they were made available only in August, though one was released one day a week to help with the tool manufacturer and to familiarize himself with the components so as to be more useful later. In Chungli, the specalist had the help of two engineers from Britain, one with 30 years experience working on layout and equipment and the other with 20 years experience as a process engineer. The specialist himself had been released from Britain only because the chairman of Ford–Europe had asked Britain to do so, on a request from FASPAC; the chairman had previously been in charge of FASPAC and had a personal interest in seeing that the complementation program in that area succeeded. But FASPAC had greater difficulty getting the other two British engineers out, because Britain was afraid it would not get them back. In addition, a British equipment engineer was obtained, part of whose task was to train engineers on the new iron furnace to be installed; it had been ordered to Ford standards in Britain. It arrived only after he left, however, and the Chinese were largely installing it on their own.

Once at the foundry, the specialist implemented the processes and trained the engineers and restarted production. He corrected the sand-moulding process, with help from a visiting British manufacturing engineer attached to FASPAC and a U.K. process engineer. He improved the melt analysis, which had been so lax that the proper hardness was not achieved; this was done with assistance from Ford-Britain and Australia. He then trained the laboratory foreman and technicians, the general foreman and the manager, the melt technicians, and the quality control supervisor; he had the technical workers visit local pressure-casting foundries to see how the work was done.

An independent company could not get this type of assistance, according to the British specialist, on any such temporary basis as was provided at Chungli. It might try to hire such engineers away from another foundry, but others (such as a New Zealand foundry) have tried and failed. A company can *buy* a foundry installation of a given type and can produce an engine out of it; but it would not have the continuing ability to improve, which results only from technical assistance coming out of the parent organization as it makes changes in processes and products that are necessary to achieve efficiency.

Adaptations were made in all four phases of production—melt, mold, core, and cleaning of the casting. These required changes in the layout, facility requirements, and equipment. The process required scaling down the experience and facilities of the larger foundries to the capacity needs of Ford-Lio Ho. The selection of processes and equipment was based on (a) what was in existence in the foundry, (b) what was compatible with the environment—that is atmosphere, quality, and skill of labor—and (c) cost.

After these changes were made and the plant facilities laid out, assistance continued during the start-up and run-in stages. One of the basic deficiencies with the Chinese engineers in the foundry was lack of diagnostic skills, which was a result of the fact that the Japanese did not teach them any analysis but merely corrected their problems. (The Japanese had kept an engineer in the foundry full-time, but he did not teach the Chinese how to "trouble-shoot.") Consequently, in one case in which the sand moulds were breaking, the Chinese had changed the sand mix by injecting more clay to make it less brittle; this was the right tactic, but (using trial and error) they also put in more of other ingredients, which raised the weight of the mould, increased the cost, and offset the effect of the clay;

the mould continued to break, therefore, until a visitor from Australia set it straight.

In another instance, the sand moulds were coming out with air pockets. The Chinese diagnosed the problem as one correctible by letting more air escape out of the joints and had filed across the vent from inside to increase its size, scratching the face of the mold. The Chinese analysis was correct, but their methods of correction were faulty, which again created additional problems. The analysis of the British engineer was that strengthening of the mould should first be tried by changing instructions to the die-cast manufacturer in Taiwan; it was done and it helped; but increased venting was still necessary. This was done, but not at the joints and without scratching the face of the mould.

Another case of misdiagnosis was one concerning the cylinder-block mould making. The mould was being broken on the jump flange face and the flange edge of the water pump—both on the same plane. The Chinese production engineer called for assistance and suggested that there was "insufficient draft angle" on the mould patterns in these two areas and that "additional taper should be added." This conclusion was made by the Chinese without any dimensional check on the patterns and toolings themselves; they simply assumed that the tooling was correct but that a change should be made in the design of the pattern. Prior experience of the foreign service specialist indicated that the cause was "bad draw on the machine" or the "flask roller heights were not parallel to the pattern plates." A check of the dimensions of the roller heights showed them out of parallel by 4 millimeters; correction stopped the breakage.

In another instance, a new, small, semi-automatic mould line had been installed for the exhaust manifold, flywheels, and bearing caps. The flywheel pattern was mounted and local personnel and engineers set up the machine. The correct method of doing so would have been to adjust the machine cycle of jolt and jolt-squeezing to accomplish the correct mould density and *then* to adjust the amount of sand dropped on the mould to ensure that the back height of the mould sand was level with back of the mould flask. But the Chinese had set it up with first regard to the mould hardness and were adjusting the cycle of the machine for sand depth and the back of the flask—omitting one important and very critical step. They apparently had not learned the significance of this procedure from Toyota, having been told merely to "do this" rather than why. (The specialist's assessment was that Toyota probably decided that if they trained a Chinese engineer, he would become valuable to another company and would leave.)

Once this specialist returns to Australia he will still be responsible for planning the next stages of expansion of the foundry, which call for raising it from 32,000 units to 60,000 units per year in stages. He will return on occasions to check progress at Chungli. Once the project is completed, he goes back to Britain. The Chinese will have to take over completely then. They are now able to assume responsibility for the maintenance of the layout and equipment by making normal improvements and replacements, and so forth. But they will not be in full production until March 1975 and will need several visits to guide their build-up until then; this assistance can be provided only from Ford-Britain, for Australian engineers are fully occupied with their own foundry problems.

To accomplish the build-up within costs constraints will tax the capabilities of the Chinese, who are not really up to such managerial decisions. They will have to get all the little bugs out and keep costs and scrappage down; both objectives require continuing surveillance, which is not their forte. Consequently, the British engineers expect foundry productivity to fall once they leave.

ENGINE PLANT

The engine plant required conversion of the machinery and retooling of 140 of the 160 machines left in the plant; the redesign of the operations and machines was done by FASPAC engineers who arranged with worldwide manufacturers to retool the existing machines. In addition, 20 new machines were designed at FASPAC to meet specific purposes with processes different from those in Britain—therefore, it could not take tooling of existing British design. Since much of the engine production is to be exported, FASPAC also had to design the packing operation.

In January 1973, a Product Engineering Manager was sent from FASPAC to Chungli for an extended stay; he brought specialists from Ford-Australia on component design, vehicle testing, and on engineering systems (drawings, filing, materials inventory and other administrative tasks). When these specialists left, they had trained three Chinese in these positions, and a replacement for the manager was brought in from Ford-South Africa, largely because South Africa was also producing the Kent engine and no one was available from Britain.

South Africa offered the principal engineer on all engines there (who had been through "engine validation" three times and was checked out in product engineering development and in testing of performance). Ford-SA was glad for him to have the experience, since he was slated upon his return for promotion to product engineering responsibility and would profit from being responsible for total performance to standards and approval of any deviation in parts or design changes.

His connection with Ford-SA has made many decisions easier, for he went back to them for the "add and delete" list on the Kent engine to help decide what parts and components Chungli should import. The engine was produced wholly out of imported parts at the time, but some local foundry castings were expected to be substituted in 1975. There has been considerable effort to obtain other components from local suppliers, and this will be done on an increasing scale, raising also the assistance needed on supply problems with the engine. He sent his recommendations to FASPAC, but they told him to proceed on his own with South Africa (which is not within FASPAC's jurisdiction) and gave him sole responsibility for determining the unique features that would be suitable only to Taiwan, which has extremely low speed limits. He decided to use the South African Cortina unchanged in Taiwan.

This engineer was trained not only in South African operations but also by a Ford-US specialist in dynamometer testing (six weeks in Dearborn) and also in testing of V-8s and in-line engines, in preparation for supervision of dynamometer testing in South Africa. He also spent two weeks in Britain on testing procedures, plus a short visit to Ford-Germany. On another occasion, he visited both Britain and Germany on engine validation, at the foundry in Britain and on tooling in Germany.

These visits were of great assistance in getting information out to Taiwan, for Ford-Lio Ho does not have a liaison man in Britain (as does South Africa). He, therefore, uses the Australian liaison engineer in Britain on the Kent engine, but has none on the vehicles. So, his contacts helped him know whom to address his request to in Britain.

If there are any unresolved problems with Ford-Britain, Ford-Lio Ho can go through FASPAC to Ford-Europe, or even directly (since the chairman of Ford-Europe is well known to managers in Asia, being formerly the head of FASPAC). Ford-Europe will then decide how to resolve the issue. The fact that some of the Chinese engineer/managers have visited operations in Britain and the United States has given them the feeling that they "have the world resources of Ford at their finger tips," as one stated in conversation.

Britain is the source of most of the specialized assistance, since it is also the source of the CKD engine components. Any supply problems in imports give rise to immediate communication back to Britain directly. Britain has also sent out from Snyder (equipment manufacturer in Britain) a mechanical and an electrical engineer to help on the equipment modification and installation.

The Snyder (and Kearney-Trecker) engineers were sent to Ford–Lio Ho to help with the machines to rough out, mill, bore, and finish the engine blocks. They were installing and checking on the machines and training the Chinese on the tool changes and maintenance. This would normally be a six weeks' task in the United Kingdom, but it takes over ten weeks in Taiwan, partly because of shipping delays. The task is made longer also by the difficulty of training the Chinese engineers and technicians, who have less experience and ability to understand the machinery, despite the fact that they are very eager to learn. Their learning consists more of watching and listening than in doing; they do not like to perform as much as to absorb by oral communication or reading. They are also reticent to take suggestions on diagnosis from a foreign engineer. They are unconcerned with *future* problems, and it is difficult to get them to be concerned about maintenance; they do not see the value of preventive maintenance, nor have they had any instruction on this in their engineering schools.

One of the difficulties of using the Japanese machines was that Toyota did not supply the engineering drawings in detail on the equipment it sold to Lio Ho. FASPAC, therefore, had to retool these machines without drawings, which required expert analysis and redrafting of specifications. Keeping 95 percent of the Japanese equipment was not the most efficient for production from a technical standpoint, but it did keep capital expenditure down and got production up rapidly. This approach required the most technical assistance and forced the FASPAC process engineer to draw on all his experience and that within FASPAC.

This FASPAC specialist, assigned to Taiwan for a ten-month period to trouble-shoot the start-up of the engine plant, was concerned with processes, plant engineering, and the tool room. His background was thirty years' experience in auto and mechanical engineering around the world, twenty-six years of which were in plant and manufacturing engineering, tool-making, and design. He has had experience in Germany, France, United Kingdom, United States, and Japan since 1970—and in *all*, language differences gave rise to problems. The problem of communication is critical in his view; one must gain the confidence of the local managers and workers so that they will ask questions, which Chinese culture makes them reluctant to do. Consequently, they may need a process or manufacturing engineer for another year or so; at present, the Chinese are slated to take over long before they would be considered prepared to do so if they were in the United Kingdom or United States.

The specialist has had to adopt for use within the plant tools that can be thrown away when worn out because regrinding of tools is poorly done in Taiwan. In addition, there is greater skill required in resetting a reground tool to get identical results as from a new tool. This expertise is not held by the Chinese at present. The main problem of this specialist, as he saw it, was to train process engineers to "think Ford systems," which have been about 90 percent adopted. It is not clear whether Ford–Lio Ho *should* be standardized to this degree with practice elsewhere in Ford, but it can add and subtract later as it matures.

It was his conclusion that no independent consultant could supply the plant or systems engineering that he could from within Ford; one has to be *inside* a manufacturing company to be able to transfer its operating practices and standards. For example, he adapted the radial drill machines from single fixtures to index-turnover fixtures with multi-tool changes. He changed the thrust-face machine, which had the tools moving out to the face, by taking out the innards of the machine, putting in a new fixture block, changing the machine cycle and feed (from *across* to *in* from both ends) to form a seat for the thrust plate. And he redesigned the flow of the engines so that after machining the engine was left at the assembly plant side instead of the other end of the plant. He added a heat-control room for pistons, then offices, and a conference room.

Quality-control training has also been critical in the engine plant. The Taiwanese have to be trained not only in testing but also in the maintenance and repair of their own guages—production personnel should not be pulled off to make such repairs. A quality-control engineer was at the plant to set up the equipment and train the Chinese, which was projected to require four weeks on the Bendix equipment and two weeks in training. Emphasis was on drawings and "show-and-tell" methods, with much more "show" than "tell" because of the lack of English at the lower job levels. After about a year of assistance from Australia, the Chinese are expected to operate on their own.

Testing facilities are critical not only to production in the line and in final phases, but also in checking CKD components: British cams, crankshafts, and pistons are not always within the tolerances prescribed, and these have to be checked out, accepted, rejected, or sent back.

Men from FASPAC and Australia are constantly making short visits; a manufacturing engineer from Australia was there for four to five months to check out machining processes, the machinery, and conveyers and to minimize scrap and rejects. Another Australian engineer was a group leader in metallurgy with fifteen years of experience. He was responsible for all maintenance of guaging in the tests on engines from the casting stage on to final assembly but was spending 85 percent of his time on training, the rest on checking and helping with inspection reports. He was also helping on production controls, though this was not really an area of his expertise. And he was helping to improvise tools and equipment in quality control because the small volume of production did not warrant expenditures on costly and sophisticated equipment. He had found that the Chinese were good improvisers and that they did not always follow Ford standards and procedures, but were often getting results; when they did so, he left them alone; if the results were not satisfactory, he would change the procedures and explain why.

The Japanese operation was 20 engines per day on a batch basis. The Ford system will produce over 80 per day on a semi-flow line basis with semi-automatic and nonautomatic machinery. The substantial difference that arises in scheduling, sequencing, and timing requires careful line-balance; some machines have only one worker at them always and others (semi-automatic) can be manned several at a time by one worker. As with the foundry, the Chinese have far to go in learning the management phases of the engine plant.

Assembly and Fiera

The assembly plant was both changed and expanded so that three models could be assembled with a quick change-over of tools—all flowing on the same line. The first to come off was the Cortina, next the Escort, and later the Fiera. In addition to the resident personnel, each model required a foreign traveller from Britain for one month to check out assembly procedures and solve any problems. But the Fiera, being a new model for Ford, required substantial assistance.

One of the first steps was to decide with Britain what Escort and Cortina parts would be imported and which would be deleted from the CKD package. (South Africa was not an appropriate source, nor Australia, for both are too expensive and are solely right-hand drive whereas Britain had the appropriate left-hand models for export). These deletions required a specialist to determine details and whether alternative materials could be used through local suppliers. If alternative materials were used, then the means of tooling and assembling these into the vehicle had to be determined. For example, scuff-plates are normally of aluminum, but these were changed to plastic, and a local supplier had to be found. The door armrest was moulded from plastic, and a local supplier was found and put into the auto business.

The plant itself was expanded and new tooling added—welding tools, timers, conveyers, and so forth and other facilities and tooling aids. The main new tools, which came from Australia and Britain, matched Ford standards; thus much technical assistance was *built into* the tools themselves. As expansion took place, assembly process engineers arrived to train the Chinese and oversee operations. Two vehicles were taken completely apart several times to permit each operator to learn more than his own operation, while learning *all* of his specialty thoroughly (e.g., *all* trimming, *all* welding, and so forth). Many workers were doing nothing but learning for some time. Those workers with Toyota background learned more quickly, but even they had to be trained on the more difficult tasks and had to be grounded in inspection and quality control.

The Fiera assembly begins with a local stamping operation in which the body panels are stamped on simple machinery; these panels are moved into the general assembly area used for all models. Technical assistance was received from five Filipines and three Australians on tooling, dies, stamping procedures, and die-changing, welding, and quality control for over two months. (The Philippines themselves had initially had the assistance of a team made up of six Australians, two Americans, and three South Africans.) The Australians had designed the vehicle but had no experience in producing it; the Philippines were a year and a half ahead of Taiwan in production.

The engineers involved could not estimate the amount of time or money saved by Ford-Lio Ho in being able to tap the design and production capabilities behind the Fiera; obviously, pilot production time was eliminated and training time greatly reduced, but there is no way to value the design costs of the Fiera that should be allocated to Taiwan. Roughly, Taiwan received over $250,000 of technical assistance in the start-up of the Fiera.

Given the necessity to expand the capacity of the assembly plant by 50 percent, FASPAC redesigned the layout of the body shop, trip shop, power and air distribution, and so forth but not the paint shop. The redesign was then approved by World Headquarters. An Australian engineer was designated as project director and sent to Taipei with twenty Australian and FASPAC personnel.

The detail and level of technical assistance that was required is illustrated by the case in which coil springs were missing from the CKD unit on Cortinas; the line began to close down at the point of insertion of the springs. The Australian resident manufacturing manager told them to put a block of wood in its place and continue production, rather than let the line go idle. When the same omission occurred later on Escort, the Chinese kept the line going but forgot to put the block of wood in.

The Chinese seem to be quite good at improvising, but only if they *see* the problem, which they may not. For example, it was decided after a market sampling to leave the fender mirrors off of cars for individuals because customers felt the mirrors made the car look like a taxi. The manufacturing manager told the Chinese plant assembly manager to delete the mirrors, but the cars came through with the fenders having the same two holes drilled for the mirrors to be attached; the plant manager replied on questioning that he had not been told to stop having the holes drilled.

The Chinese manager of the assembly plant stated that the most important lessons learned from the Ford specialists were that (a) if a breakdown should occur along the line, to shift workers from elsewhere to fix it, or shift the production workers elsewhere for a time; (b) if spare parts become scarce, send someone out to get them expedited rather than just waiting and letting the line close down. He considered that they would probably have more trouble with getting British CKD units after the Ford specialists left than they would in local operations; there was simply too much damage in the CKD units and too many parts missing (even compared to those coming from Toyota). He felt the only way to fix this would be to take a trip to Britain to establish personal relations.

The paint job in Taiwan is probably better than in areas outside of Southeast Asia because it is feasible to hire more workers in this shop per car for inspection and correction of errors. And fewer mistakes in production tend to be made because the line moves slower (by manual pulling of the dollies in most cases) and workers do more different tasks within a given area and therefore feel a greater sense of achievement than if they were highly specialized.

One development that was tested out in the Chungli plant to improve assembly operations was the low-volume pivoting pillar buck (PPB). This equipment was designed by Ford–Australia and FASPAC's manufacturing staff in order to reduce costs, speed welding, handle a variety of models, and reduce down-time during the change-overs and reduce floor space. The PPB substitutes for the conventional "post and latches buck" and permits better access by tipping the body from side to side to facilitate welding. The tests on the PPB were so successful that Ford–Lio Ho is now making the equipment and it has been proposed for adoption in the low-volume plants in

New Zealand, Philippines, Thailand, and Malaysia for the new Brenda model as well as the Cortina. Taiwan will make parts of this equipment for supply to the other affiliates, thereby increasing its export volume.

Quality control remains a difficult problem for Ford-Lio Ho because few of the former employees were trained in this area. The Chinese did have a quality-control officer in the days with Toyota; but in order to meet Ford standards, the new manufacturing manager took over on his arrival from Australia. He introduced inspection forms for checking as the vehicle passes through the system, tests on supplier parts on engineering standards and size before they are brought to the production line, sample tests in the plant, and in-process inspection of such tasks as spot welding (strength of weld), surface finishing, size of window and door openings, sealers, bond-erite solution, water, heat for paint, paint (scratches, thickness, marring, and so forth), interior materials, and so forth.

Training on quality control is given through lectures by the manufacturing manager, through regular industrial relations programs, and by the Chinese manager who was trained with Toyota; further training is on the job. To improve skills in this area, one quality-control auditor was sent to Australia to study a new system for a Uniform Quality Inspection Audit recently introduced by Ford that would permit uniform determination of the quality of the same vehicles all over the world; it was taught in Australia by American technicians.

The type of technical assistance available on these assembly problems is simply not available within Taiwan nor from international consultants—in fact, no such capacity exists outside the industry itself. This makes it impossible to value the assistance by comparing costs of different sources of supply. The magnitude is best seen in terms of the fact that 350 man-months were required from the point of program planning to launch from outside of Ford Lio Ho. These were charged to Taiwan; what could not be charged was the donation of FASPAC time in conception of the project, requiring 18 man-months, and the development of the package for assessment by Dearborn, another 24 man-months. In addition, there was an estimated 50 man-months of support by FASPAC in Australia, not charged to the affiliate, plus 60 man-months from other functional areas outside of manufacturing. The estimated cost per man-month is between $4000 and $5000. Conservatively, the total of nearly 500 man-months would have cost some $2.5 million, equal about 18 percent of a total investment of U.S. $6 million in purchase of 70 percent of Lio Ho and $7-9 million for new fixed assets (tooling, etc.) and for related launching and engineering support.

TRAINING AND RECEPTIVITY

Training needs at Ford-Lio Ho are critically affected by four factors: one is the fact that Toyota had not taught the "why" of actions desired; another is the youth and inexperience of the staff, from Chinese managers down; third is an annualized rate of turnover of 36 percent, mostly among new employees; and fourth is the loss of certain key engineers and machinists during the change-over from Toyota to Ford models. The youth of the staff is explained in part by the practice of Lio Ho owners of training technicians in their own school for various industrial activities; these students were at high school age in the main. The loss of so many employees in the interim period was recouped a bit by the hiring back of some of the workers when operations were begun. Those newly hired are also young but are well-enough trained and can learn readily even without English capabilities; in fact, English knowledge does not correlate necessarily with good performance in the plant. Given the inexperience, however, efficiency in the plant is at 70 to 75 percent of what is sought, and it must reach at least 90 percent to be acceptable.

This lack of experience and application is reflected in the fact that the company had to develop its own tool room for tools that it could not obtain locally. There are simply no patternmakers in Taiwan, so they had to train their own men at the site; this is very difficult and is usually a five-year task.

Despite the youth and inexperience of the Lio Ho managers, a few have had considerable experience. One individual who is a likely candidate for the top post of manufacturing manager was trained as an aircraft engineer; spent six years with the Yue Loong auto company; was manager of manufacturing engineering at Philco-Ford, spending two months in Philadelphia; later was with Timex; was a division manager in manufacturing engineering and plant engineering with Zenith; and finally was a program manager with Ford-Lio Ho. In addition, he is a part-time professor in industrial engineering and manufacturing measurement at one university and also part-time at another on production control; he considers most such training as too theoretical and in need of more applications. He graduated from a mainland China university in mechanical engineering and was a graduate in aero-engineering on mainland China. He is also a senior member of the American Institute of Industrial Engineers.

The Chinese are eager to learn and will seek out learning situations by asking for new opportunities; some will even change the methods of operation simply to experiment and see what will happen, without any other "reason" for doing so. Training is done by everyone— foremen, "leading worker" in a group, industrial relations personnel, supervisors, managers, et al. As a consequence, 70 percent of the workers learn more than one task. However, their application of this training is directly correlated with their willingness to perform— that is, to practice rather than merely learn. Training on-the-job, therefore, is often a task of getting workers "to do," to check for mistakes, to question how they can correct, and to examine both diagnosis and correction and explain why a proposed solution would or would not work.

Eagerness to learn is usually translated into readiness to read books, but not to practice on the site. The Chinese will seek promotion through education, which is but one of four criteria used by Ford-Lio Ho; the others are skills, personality, and leadership (trust, correct thinking, ability to train, and common sense).

To assist affiliates such as that in Taiwan, a "Planning Guide for Training" has been prepared by World Headquarters; based on the Canadian experience, it runs to 200 pages. An annual training report goes to FASPAC, with whom there is a discussion of problems and deficiencies and how to resolve them. These exchanges are supplemented by regional meetings of industrial relations officers every two or three years and by visits to other affiliates off and on during the year.

Training Programs

Company training programs are encouraged by a governmental requirement of contributions of 1.5 percent of salaries and wages to the Vocational Training Fund Board; if the company has a training program of its own, it can recover 80 percent of this contribution. To emphasize training, all plant managers, supervisors, foremen are also designated as "instructors." A training section has been set up within industrial relations, which uses a two-room training center in the administration building. Four distinct training programs have been established: one for new employees consists of a four-to eight-hour induction course introducing the workers to the facilities, safety, functions, layout, and so forth; a second involves in-plant, "on-the-job" training each week for three and a half days in each of the major areas plus quality control; a third provides supervisor training; and a fourth relates to off-site training and education.

On-the-job training is not simply instruction at the machine; it involves selection of 12 workers in the foundry, 16 in the engine plant, and 22 in the assembly plant for a set of courses meeting each week for several weeks. Two sets of students are trained each year; a certificate of completion is awarded and records are kept of the results and placed in the employees file. In addition, about two-thirds of present machinists in the engine plant are apprentices; considerable "show-and-tell" training is required before they become proficient. Other examples of on-the-job training include two engineers in the engine plant: one works 60 percent of the time on testing but also on assembly and validation; the other, on design changes coming through Britain on supply parts and castings plus testing tolerances. All expatriate personnel are required to train their own replacement and keep records of his progress, still another is being trained to replace the one moving up.

Although on-the-job training of workers predominates, a classroom approach was required in the introduction of "core-wash dipping," instead of painting the wash on. This teaching has been the exception, however, and is not generally used simply because of the time pressure to get the plant in operation.

Supervisor training also employs classroom work and provides 24 hours of instruction in leadership, safety, productivity, production methods, and so forth. It is given to group leaders, "lead workers," and foremen. It is offered twice a year to selected individuals. At present there is no training above these levels in the plant, but the company anticipates introducing courses on supervision, inspection, quality control, and so forth when it can obtain proper instructors. Over 350 employees have received formal in-plant training.

Personnel have been also assigned to one-day or one-week courses in various technical institutes, as well as to English-language training. A total of 242 employees have received outside training. And, of course, the Lio Ho Institute is a continuing source of technical training.

At a still higher level, the company has sent seven foremen and supervisors to college with full expenses paid since evening classes are nearby. This support program will be expanded to include completion of high school and work loads will be shifted to permit study time primarily by avoiding overtime for these workers.

The present educational status of the employees includes seven with masters degrees, 147 with college degrees, 281 with high school diplomas, 266 with junior high school education, and 200 with only primary education.

At present Ford-Lio Ho has not made any scholarships available nor support grants to institutes, but it has already begun dialogues with several looking to such programs. It also has brought six students into the company during the summer—two in produce development, two in manufacturing, and two in finance—from an undergraduate program, paying them each $40 per month. Each Saturday, the training administrator meets with them to discuss their experience, impressions, and so forth. One anticipation, of course, is that they might become future employees.

In addition to these programs, overseas training is provided top-level officials and technicians. Seven individuals have already been sent to various countries abroad: one to the Philippines and Australia to study industrial relations for one and a half months; one to Australia to study quality control audit; two to Malaya; two to the United Kingdom on the new Brenda model; and two to South Africa and Australia for plant engineering and product management. This program has just begun and will be supplemented by top-level management courses once sufficient time and materials are available. Within this program the key element must be education to decision making and initiative, for formal education in China is bent toward *obedience*, which does not lead to independent judgments. On the contrary, the Taiwanese are not taught to think but to memorize, repeat, and maintain order and discipline. Aggressiveness is identical to selfishness, and acquisitiveness or "attack" is considered anti-social. Individual relationships reflect the order of the social system and one would not disturb it by offering suggestions that would imply a superior was not doing his job well.

In contrast to the Japanese, who took seven top men to the Toyota operation, no visits were initially made to Australia by Ford–Lio Ho employees; rather the expatriates came to Chungli. The reason was the sizable difference between what was in Taiwan and the processes in Australia, for Taiwan is 20 to 30 years behind the United States and 15 to 20 behind Australia in processes and sophistication of machinery. Also, foreign officials have less time to spend helping if they are still on their job in their own affiliate company. A few Chinese later went to Australia for training, however; for example, the chief quality control inspector went for five weeks to review procedures in foundry operations, engine machining and assembly, and body assembly.

In 1973, one of the product engineers was sent to Dearborn for several months to work in a variety of operations (personnel and organization, product development, engines, assembly, sales and marketing, finance, customer services, and export) to prepare him for future advancement. Similar preparation has been given some of the Ford–Australia personnel, some of whom, in turn, have assisted at Lio Ho; for example, one job analyst in the Industrial Relations Office at Melborne was sent for eight months to Dearborn, where he spent three and a half months in personnel and organization, three in training services, two weeks in assembly, and two in staff planning.

Several Chinese managers suggested that if there were changes in the engine or new models, it would be good training for the product-engineering manager to visit the British operation, and probably the senior engineer and a supervisor or two. Also, if Taiwan wanted to design a new part or component, it should send an engineer to the Ford–Australia design center; but even that might not be sufficient, they asserted, for engineers there do not have time to "lecture" a neophyte. Plans were being made in 1974 for such visits to Australia by a tool analyst and a preventive maintenance engineer.

Receptivity

There are a variety of problems related to technology transfers, not the least of which relate to communication, availability of basic learning skills and education, and willingness to accept assistance. All three are key elements of the culture of the host country, which may be constituted so as to reject some of the facets of industrialization that U.S. companies take for granted as necessary.

The problem of communication is most readily seen in the differences in language, which all officials in Taiwan noted as a major obstacle even when the Chinese spoke some English and which can be illustrated by the incredulous reaction of the Chinese engineer to the Australian resident who had just driven his car in to be checked because "the motor was missing." If the Taiwanese do not understand English instructions, it is from lack of comprehension of the idea because of great differences in mode of expression. Since the language itself involves repetition, questions and sentences must be rephrased several ways to elicit full answers; otherwise, one is likely to get a positive response when the Taiwanese listener means only that he has heard—not necessarily that he has understood or will act.

The Chinese who reach higher levels of responsibility do speak English, of necessity, for most of the paperwork has to be in that language in order to transact business with Australia, FASPAC, and the British supplier of CKD units. In addition, all instructions come in that language. Translations are made on the spot to assist understanding by those lower down in the operation.

But even if English is spoken, it does not mean that the Chinese *think* in English, which is a more rationally oriented language than Chinese. Language communicates not only words but feelings and relationships (all Chinese are *Mr.* . . .) through the words chosen, those *not* chosen, the emphasis given them, accompanying movements or facial expressions, the setting in which they are used, and so forth. This "silent" communication is affected by a wider understanding (lack of understanding or misunderstanding) of the motives and intent of the other party. Misunderstanding arises also from ignorance of the cultural biases of the other; this gap can be closed only with continuing contact, visitation, and greater familiarity. The *physical* presence of the foreign specialists, therefore, has been critical to the transfers into Taiwan; and visits by Chinese to Australia, Britain, and South Africa have been most helpful in narrowing the gap, dissipating distrust, and reducing the tendency to "play games" through the Telexes or written communications. Any reluctance on the part of a foreign affiliate, such as Germany, to extend assistance is magnified at the receiving end, especially in Taiwan where feelings of national (cultural) pride are high. The gap is somewhat narrowed by the existence of a liaison engineer in the assisting company—as South Africa and Australia have in Britain. Ford-Lio Ho does not have such an engineer there, and that function is performed solely by a British KD manager responsible for a particular affiliate's orders and continuing satisfaction.

Lack of cultural understanding and the resulting distrust magnify the feeling that certain information cannot be passed to an affiliate; where this distance is increased by having a mere licensing agreement rather than a controlling equity interest, much information simply will not be given. The view that information that is proprietary to the company might be leaked out or given to the host government, if in the national interest, makes the donor cautious and concerned over the "loyalty" of the recipient. (It seems likely that members of a culture such as that in South Africa will be seen as "more loyal" than those in Taiwan simply because of the cultural gap, which may or may not reflect the real loyalties felt.)

The perceptions of the recipient company are, of course, equally significant, especially when they are in error. It was reported several times that the Chinese see Ford as having plenty of money and therefore able to buy all the new equipment and advanced facilities Ford-Lio Ho would want, and the Chinese want the *best*. Many of the foreign service specialists asserted that it was difficult to eradicate these views on the part of the Taiwanese and that a crash course in Taiwanese culture and language might have helped in establishing closer communication and therefore improved the background for technology transfers.

A lack of understanding can create many difficulties—not only in adding to distrust and lack of credibility but also to confusion and therefore greater defects, lower productivity, increased rejects, and "cripples" or deficient vehicles stored in the yard. The basic lack of an understanding of the necessity to schedule, to plan, to program, and to organize a flow of components and products is itself a reflection of a concept of *order* that is more static and imposed from outside than dynamic and created from within the group. However, this same concept of order facilitates the carrying out of direct orders from a superior; instructions will be carried out to the letter, but the implementation also will not be modified even if it creates a further problem or if errors arise.

One of the greatest difficulties in transferring technology to Ford-Lio Ho, therefore, is that of inseminating a *systematic* approach with check points, tests, modifications, "buy-outs," and changes in procedures to eliminate problems or work around them. It is difficult to get the Chinese to understand *why* such an approach is desirable, as compared with simply proceeding according to the rule book.

Ranking close to this difficulty is that of transferring the concept of working to budget or costs; money is seen as no constraint since the company is rich; or if money does run out, a modification will be made. Improvisation occurs readily, but only by necessity— not by plan. The idea of budgeting to put the money where it does the most good is foreign and difficult to accept. One reason for this is the extreme reluctance to alter (or disregard) the standard procedure or quality requirements on a given operation or component; they do not have the confidence that comes from experience to make an on-the-spot decision as to whether a deviation is permissible *ad interim.* (Yet, on occasion, it appears that everything breaks loose and all rules are violated with abandon, with very strange decisions being made or actions taken with no apparent reason, such as when the air circulation was turned off on a sophisticated machine, thereby causing it to heat up.)

The Chinese also seem unable to anticipate what *might* happen if a particular act is not taken within a sequence they have been told to follow. Thus, there is no preparation for breakdowns—no contingency planning—so, when the paint shop broke down, the entire facility ground to a halt, and no one sought ways of keeping going because the paint shop "would be fixed in a few hours" (it wasn't) and everything could then go on as before. In another instance, a test driver lost a wheel from a test car because he did not tighten the bolts on the wheel after changing a flat on the road. The Chinese need a variety of experiences to learn from, but they do not always learn from bad experience; still, this does not worry tnem or their engineer/managers, for they know they will eventually succeed, and the *time* in which it is done is not so important! Their concept of time is itself a factor in their willingness to accept assistance, for it reduces their willingness to concern themselves with scheduling.

As to education prior to entering the company, the levels of preparation tend to be quite high, as illustrated earlier, and the caliber of university training is high, but it is not slanted toward applications. Therefore, engineers do not always understand what they are doing when operating on the floor—they perform by rote, for that has been the way they were taught. Consequently, they are not always able to describe what they are doing or why.

The absence of applied education interferes with on-the-job training, which appears as simply taking orders. This position also rankles, so they will sometimes simply change a procedure or process in order not to be merely "copying"—a posture they reject. It reminds them of being treated as servants and was difficult for them to take under the Japanese arrangement; consequently, that operation was at 40 percent of expected productivity. With some authority and freedom under Ford, which treats them as equals, they feel more comfortable but sometimes exercise it without examining the consequences of their acts.

The foreign specialists expected to find a higher level of technical expertise at Lio Ho than they did, because of its prior experience with Toyota. But Toyota had not transferred any high-level technology, despite the fact that the Taiwanese foreman was in Japan for ten months before the opening of Ford–Lio Ho. Consequently, many of the Taiwanese thought they were trained and could not understand the gap between their capabilities under Toyota and what Ford was requiring. They need training on simple techniques of adjusting to breakdowns of gaps in supply or repairs to vehicle in the line. For example, if a car did not pass an in-line test, the line would be stopped until someone came to fix it; or, it would be trundled back into the line at the point of operation, which would drop productivity by 75 percent through interruption of the sequence. Similarly, during quality-control inspections, if a part was missing, the line would be stopped until it was found. Or, if the vehicle came to the finished inspection and a part was missing, it would be put out into the yard for eventual completion but not recorded as "produced" during that day; whereas, it should have been recorded as "built, but short."

On the side of the donor company, officials of Ford–Lio Ho stressed the need for adequate education of foreign service specialists.

Some have been assigned who were not capable of making the transition into the local conditions and were not able to teach effectively; and some so quickly that there was not time for cultural adaptation. The most critical quality of effective teaching of the Chinese is *patience;* the second is ability to transmit ideas in understandable patterns; and the third is willingness to write everything down. The written word is "sacred" while oral words are soon forgotten; written are mandatory and permanent and will not be changed while oral are permissive and passing. This attitude is generated during their education, which is based on books.

If an error is found in a process, the Chinese will seek to find the new rule by trial and error, rather than by analysis, thereby digging a deeper hole and reducing their chances of finding a solution by 90 percent, according to the foreign service specialists. This means a lack of ability to "trouble-shoot." In their "trouble-shooting," the Chinese will always take the easy road first, and will therefore often have to repeat and repeat to get it right. Even in facing choices of what to do next, they will take the easiest first, rather than undertaking operations in sequence, because they may not *have* to do the more difficult—who knows that tomorrow brings? Consequently, when the specialists leave, it is expected that the Chinese will have trouble with several machines and will require visits back from FASPAC.

A final problem in transfers of technology is that of the willingness on the part of the recipient to accept instruction. The Chinese have a strong desire to try it "their way" first and, therefore, will attempt to re-invent the wheel before they will finally accept the procedure recommended by Ford. This trait exacerbates another difficulty arising from the fact that they are very sensitive to *personal* criticism and do not like to lose face by making errors. Therefore, when an error is made, it is not personalized—the machine is blamed or the system and procedure, but not the individual—thereby reducing the necessity for the individual to change his behavior (i.e., to learn from doing).

The sensitivity to instruction is illustrated by the engineer who was sent to South Africa for a month's training, learned dynamometer testing of the engine, returned to set up the facility, and trained others, but left after six months because he was insulted to have to take assistance from Australians at Lio Ho.

In quality control, the Japanese serviced the guaging of the machines only once a year. Ford requires constant checking, but this means that the guages may be wrong at times and decisions have to be made as to whether the products are acceptable or should be rechecked. This goes against the Chinese desire to have "go or no-go" guages, which do not require personal decisions and permit faulting of the machine rather than the operator.

At the same time, they will often take an initiative to change or to seek a new route. For example, the Chinese buyers for the company are not sufficiently trained in analysis of parts or how they might be improved by the supplier to help the company, and they will hesitate to try to change the supplier or bargain strongly on price (again, the attitude is that "Ford can pay!"). But they *will* follow explicit instructions to "seek a supply of glass at 10 percent price reduction" and will even suggest alternative ways of doing so.

SUPPLIER ASSISTANCE

Ford-Lio Ho does not face the same local-content requirements that are found in South Africa, but they do necessitate a 40 percent Taiwanese content by value. The company has, therefore, had the assistance of FASPAC personnel. By mid-1974, some 200 separate items were purchased locally for the Escort and 210 for the Cortina.

Most of the local suppliers are licensed by Japanese companies; for example, those in batteries, horns, leaf springs, wire harnesses, blass and tires. A U.S. company also licenses another tire manufacturer, and several companies—carpets, paints, and headlamps—are self-sufficient in technology.

A complex supply problem arises with the Fiera, for it is part of a complementation arrangement in which several affiliates will supply each other, and a supplier in one country will produce for all. Common purchasing arrangements must, therefore, be made by FASPAC; each affiliate then places its orders within the overall agreement. Thus, Japan is the source for steering gears, windshield wipers, shock absorbers, and steel sheets (to be stamped locally). Argentina supplies light rear axles and differentials (from a U.S. subsidiary there); Ford-Britain supplies the engine, transmissions, and brakes; and Ford-Australia acts as an intermediary in the purchase of the front axles, spindles, some heavy axles, and miscellaneous parts from independent suppliers. This intermediary service helps to maintain quality, keep suppliers on schedule, and reduce freight costs—for a fee to Ford-Australia. Some 150 parts are obtained locally, such as windshields and wipers, lights, tires, switch indicator, horn, tools, mirrors, floor mats, and mufflers.

All suppliers are given the Ford Supplier Quality Assurance Program, translated into Chinese. To the extent that it is followed, other auto companies benefit from the improved quality and scheduling. The government imposes tight quality controls on the companies and is quite interested itself in this program.

The potential for local supplies of axles exists, with those for the Escort and Cortina coming from the Dana company, which purchased a Taiwanese company. The actual local content in local supplies is difficult to determine, for the total value of an item purchased from a local company is counted, even if it contains substantial import content; for example, coil springs are imported by a seat manufacturer, but they are counted in the final value as "local." Even the import of a complete part of an "exclusive local importer" is counted as local content; but Ford could not itself do the importing.

Technical assistance was given to Ford-Lio Ho on how to carry out its "jobbing" operations and how to estimate production costs of the supplier. In turn, suppliers have been extended assistance on such items as seats and door panels. In one instance, the company redesigned the rear seat of the Escort, which is used as a taxi in Taiwan (but nowhere else in the world because of its small size) and assisted the supplier of seat frames to make the changes. There has been no need to extend assistance on items such as batteries (which are a shelf item) or mufflers.

The major difficulties with suppliers have been in their inability to meet Ford standards and in poor delivery scheduling. Consequently, compromises have had to be made on design, tolerance, or even materials in some cases; a high rejection rate immediately elicits assistance. Ford-Lio Ho has had to change the bracket for the horn, the carpet, and some minor items: for example, from molded carpets to "cut and sew," which suits local techniques. Two expediters are on the road constantly trying to accelerate scheduling and delivery among the handful of suppliers that are habitually late.

An Australian expert in purchasing came for six months in late 1973, finally found some new sources in the south of the island, and was needed for an additional six months. Many companies were introduced into the industry by adaptation of existing operations; for example, a maker of refrigerator bodies into other light stampings. Over half of the suppliers made such adaptations; the rest were already in the product line. Ford also introduced several companies to potential U.S. licensors of components.

The types of assistance provided to suppliers include packaging of parts sent to Ford-Lio Ho, heat treatment, and quality control, plus some new techniques for the seat springs and polyurethane cushions. Ford has also supplied tooling to some companies on loan. No effort has yet been made to assist them in their own materials purchases or in cost reduction. But the company has had to become firm with some suppliers on the prices they have asked, for they have tried to take advantage of Ford as a "rich foreign company" and charge prices higher than those to competitors. In one case, the ability to find an alternative supplier has cut prices substantially.

The greatest need in the supplier field is to find competitive sources to reduce monopoly pricing; high local prices mean considerable penalties in cost of the vehicle not only directly but relatively, for the local-content requirement is based not on the local price relative to selling price but the pound-sterling price of the deleted items as a percentage of total value of the vehicle in pounds. Thus, the CKD unit has deleted 40 percent of the value of all items in an auto, but the local price of producing these items may be 200 or 300 percent higher, and some items have a cost penalty as high as five times the British price of the item. Ford-Lio Ho would like to reduce the average penalty to no greater than 50 percent (or a local price no higher than 150 percent of the British price). One hopeful sign is that there are numerous requests by companies now seeking to supply, whereas earlier Ford-Lio Ho had to seek out companies and show them how to produce particular items.

ASSISTANCE IN SERVICE AND REPAIR

Ford-Lio Ho extends technical assistance to dealers through its Ford Marketing Institute (FMI) on general service development. Ford-Britain and other affiliates will send out teams of trainers when a new model is adopted in any country to help train sales and service personnel. Video tapes are sent out on maintenance and repair, with a second audio track for translations to be dubbed in. These will be shown to maintenance mechanics who are sent in by the dealers or garages to a scheduled showing. Cost to Ford-Lio Ho for the equipment to show the tapes is $3,000, but well worth it.

Instruction on parts supplies and inventory control is obtained from Ford-Australia personnel, who in turn are trained by U.S. technicians sent out to FASPAC. This instruction is passed on to each affiliate four times each year. Dealer training is backed up by a Parts Distribution Center, which has been set up to Ford standards and has locals trained to man it. This support is considered so significant that an estimated loss of over $1.5 million would have occurred if the launch had been attempted without expatriate assistance. After launch, six different visits of a week or more were made by expatriate personnel on parts supply to dealers.

In addition, British trainers are sent out three or four times each year to visit with Taiwanese dealers and to instruct mechanics. There are four or five such trainers in a department in Ford-Britain with responsibility for the FASPAC region; the same individual will return to Taiwan, thereby getting familiar with the people and their problems from year to year. Ford-Lio Ho translates and prints the drawings and hand-outs he brings with him. It also helps select the individuals who are to attend particular classes he may hold with emphasis first on the service managers. These courses will be repeated each year and run for four to six days, with 20 to 25 in each session or roughly two from each distributor at present. The content includes service management, gear box repair, Escort servicing, Cortina servicing, electrical systems, diagnosis, and Fiera servicing.

In addition, Ford-Lio Ho personnel conduct courses. Technical experts in such items as brakes, differentials, engine, body, clutch, electrical systems, drive shaft, steering, and so forth on each model train garage mechanics at the Chungli plant. These are repeated once every two months or so, with two mechanics from each dealership in attendance. Since the turnover of mechanics is low (not much among dealers or between them and the plant), this program should get to all mechanics in a short time.

Three service personnel, based in the Taipei office go out on visits to dealers' service shops every day, except when conducting courses at Chungli or Taipei. They will get to each shop every week or two, mostly doing "trouble-shooting," a technique with which the Chinese are not familiar.

Two office representatives and one manager on parts supply also help the dealers with their problems of supply and inventory. Ford-Lio Ho has its own computer program on parts inventory and supply, but it has not extended the facility to dealers as yet, for their volumes are not big enough.

Ford-Lio Ho also extends technical assistance to the military on trucks that are built by the military with 60 percent local content and the rest imported from the United States. The government wants 100 percent local content in all U.S.-designed parts; it has asked Ford-Lio Ho to make these parts—castings and machining of engine parts, especially to help in rebuilding U.S. motors. Ford does not now have the capacity to do so, but it sends engineers and machinists over on occasion to help out.

Most of this instruction comes out of a plethora of manuals and training guides supplied by Dearborn, FASPAC, and the European affiliates. Manuals from World Headquarters cover the setting up of a repair shop, parts department, lighting, power requirements, tooling, and so forth. There is more along this line than can ever be used in Taiwan, including servicing of specific parts. Ford–Britain sends manuals on strip-down procedures and assembly of the V-8 engine and the 6-speed transmission. Ford–Germany sends workshop manuals on the Cortina, which require translation (which in itself is a great learning tool for the Ford–Lio Ho engineers who have to do it). Because of these ties, when there is a specific problem not covered in the manuals, Britain or Germany can be asked for assistance. For example, one set of speedometers produced by a local supplier ran 10 to 15 mph out of order; one was sent to Australia for a check out; they found the error and sent some of their own spares to keep production going. The parts could be readily used because of the *standardization* of components on any given model, which is a technique of considerable assistance to an affiliate.

DISSEMINATION

Despite the fact that there are relatively few exciting employment opportunities compared to the number of technically trained employees throughout Taiwan and that general unemployment remains high, there is considerable turnover in the lower grades of jobs and, therefore, a potential dissemination of skills out into the economy. In the first eight months of 1974, 130 employees left, of which over half had been employed less than nine months; three-fourths were in the assembly plant, and over one-fourth were "terminated for cause" and a sixth left for military service; a tenth left to further education, and another tenth for health or family reasons. This leaves only 37, or one-fourth of those separating, to go to better jobs or with their own business, thereby potentially disseminating skills into the wider community.

At higher levels, some dissemination has occurred. For example, one employee in the casting operation learned how to test for hardness; he left to start a little shop on his own, casting copies of small mechanical parts not readily found in the stores around the city. Many of the mechanics and spray painters in the plant moonlight their skills with repair shops, particularly on Sunday; but a large number of the men workers live in the dormitories provided by the company on the plant grounds and, therefore, do not have the opportunity to go into the city for regular night jobs. One mechanic in a dealer's service shop, however, became so skilled at all aspects of service and maintenance that when the owner wished to diversify into other activities, he was taken on as a partner.

Turnover among engineers is very low at present, for they are all interested in learning and are still learning a great deal. Once they consider that they have learned all Ford has to offer and are merely practicing their knowledge, some may look for new activities to learn and move into other industries. Whether this will create a dissemination of technology will depend on how they approach their new tasks.

✳ *Part II*

International Telephone and Telegraph Company

Chapter 5

ITT-South Africa: Standard Telephone and Cable

The production of private telephone switchboards (PABXs) in South Africa by ITT's affiliate arose out of transfers of technology in the field of transmission lines and telecommunications. PABXs are private automatic branch exchanges and are produced by a Division of STC-SA responsible for Business Communications.[a] This division was added to STC-SA as a result of successful expansion of the transmission sales and distribution of imported products from other ITT affiliates.

The British STC became affiliated with ITT in 1925 and established a branch office on South Africa the same year. Over the years, many sophisticated systems were produced by STC for the South African Post Office: a 12-channel for transmission after World War II, an underground cable system in 1950 allowing 600 channels, and a microwave radio network with 7,200 speech channels transmitted and received from a single antenna.

In 1948, telecommunications equipment was assembled in South Africa by STC's affiliate, with the important effect of training engineers, technicians, and production operators in transmission equipment assembly. STC-SA was incorporated in 1956 and in 1958 was accepted as the exclusive supplier of transmission equipment to the South African Post Office (SAPO) for the purpose of

[a]*Private* means they are owned by the user and installed in his office; automatic means that there is no *hand* corrections of the outgoing calls; *branch* means directly the local extension phones (as few as 5 and up to 8,000) are tied into the main telephone system. These exchanges can also connect extensions *within* the branch exchange. A *PAX* is only for calls between extension phones and not to the outside; a *PBX* is a nonautomatic exchange that is tied to the main system.

developing a self-sufficient source. The agreement required establishment of production facilities; creation of supplies of components including quartz crystals, vibrators, transistors, and so forth; and provision of comprehensive technical services on system planning, design, and application. The annual value of manufactures increased 15 times in six years and local content has risen to 80 percent.

STC–SA has become completely self-sufficient in manufacture of infeed requirements in quartz crystals, silicon transistors, transmission coils and transformers, and silvered mica, paper, and polystyrene capacitors. The knowledge gained in these fields has put STC–SA in a position to pioneer in South Africa the manufacture of micro-miniaturised electronic circuits, which will become a feature of many analogue and digital equipments in the future.

These developments were possible only because the parent company put at the disposal of STC–SA its extensive research, development, and manufacturing resources, thereby ensuring effective training of professional and operating personnel. Training was done in both Boksburg and in London, with the result that only five employees in the company have been sent in from outside.

The underlying motive of the South African government for local production has been security; it has insisted on local availability of production know-how and technical resources in all aspects of electronic component manufacture and usage from basic research in materials to specific end-device applications.

STC–SA's success in these developments and in fulfilling the 1958 Agreement with the Post Office are the background against which the entry into PABXs is set, for the Post Office must approve all switching systems used by private institutions in the country and therefore has a significant role in new product development.

HISTORY

Prior to 1963, PABXs were imported from Europe through several competing companies (STC, GEC–AEI, Philips, Plessey, and Siemens). STC assembled the Stepmaster; it was available in two models: one of 10-junctures/50-extensions/6-links, or one of 15-junctures/100-extensions/10-links. Both had two-motion selectors—that is, "step-by-step" or one step over and one step up (or down) until the crosspoint required was reached to make the connection. In the early 1960s, SAPO designed a rotary uni-selector system, which it asked the companies to build. All four companies responded by building to the specifications, designated as SAPO I and SAPO II; the former was a 3/9/3 model and the latter a 5/20/4. These models were sold by STC–SA until 1968, when the Stantomat series was introduced, along with the Pantomat 10.

The Pantomat 10 was an all-relay model with 10 extensions and 3 junctures and links (3/10/3). The Stantomats used a rotary uniselector, with models providing 5, 10, 20, 25, or 40 extensions. The Pantomat-10 and all but two of the Stantomats are obsolete now, with the last Stantomats being discontinued December 1974. All are being replaced by the new Pentomat series that is based on a cross-bar switching. Both the Stantomat and Pentomat series were taken over from other ITT affiliates, with many piece-parts imported into South Africa from sister companies. Growth of operations is reflected in a 4,500 percent increase in sales over the decade since the first year of the SAPO series (R73,000 in 1963 to R3,212,000 in 1973).[b] Company sales compare to a five-fold increase in the total number of PABXs in use between 1963 and 1973, with a sixty-fold increase in number of extensions and a two-fold increase in selling price per extension.

This growth in demand took a decided jump in 1969 after the introduction by the Post Office of a fully operational National Subscriber Trunk Dialing Network. Business organizations needed trunk barring equipment, and private subscribers needed private metering facilities—both of which could be provided by a PABX. Demand jumped sharply again in 1973 with the introduction in 1972 of high-volume extension systems that permitted several hundred extensions in the same PABX, which was needed because of the rapid expansion of industry and commerce in the country. The growth of total demand has averaged 28 percent per year for the past several years with STC–SA's sales increasing 10 times in the past five years—much too fast for mid-management in the company to absorb and deal with adequately. STC–SA and Philips are the market leaders, with between 25 and 30 percent each.

The largest number of PABXs are in the low-extension models—from 1 to 9 extensions and 10 to 25 extensions—but the largest number of extensions result from the high-volume models of over 100 extensions each. Plessey remains dominant in the small models; Siemens in the mid-range of 26 to 100 extensions; and Philips has been dominant in the over 100 models, though STC–SA has been second in each series and therefore at the top overall. Demand for new equipment is expected to taper off in the future, with replacement demand accounting for 40 percent of sales, at least in the middle range.

[b]R Stands for Rands, roughly equal $1.50.

ORGANIZATION

The Business Communications Division of STC–SA was originally a distributor of Chrysler air-conditioning equipment under an agreement between ITT and Chrysler. It has since dropped this line but still is a substantial distributor of a variety of business communication equipment, including cables, voice peripherals attached to phones, postal automation equipment, internal voice systems (PALX), private automatic exchanges (PAX) for *within* a company, pocket-paging devices, and mobile radio equipment. It has examined, but rejected, industrial temperature controls and is still looking into new products for distribution.

BCD's major manufacturing effort is the PABX. Other manufacturing divisions within STC–SA include the Telecommunications Division and Components Division.

The managing director of STC–SA is responsible for these, plus a wholesaling division, and has a staff dealing with personnel, technical, finance, operations, and contracts. The technical staff is composed of a director (engineer), plus one on switching and another on telecommunications. The BCD is comprised of units responsible for manufacturing, marketing, engineering, control and testing, and market research. The last is looking for new products to distribute or manufacture; the rest have to do with manufacture of PABXs and are located at the manufacturing site in Boksburg, just outside Johannesburg, along with the transmission and telecommunications operations.

As to STC–SA's relationship to the parent company and the rest of ITT, a company brochure states that a "new status" has been given the SA operations: "Now recognized as a fully autonomous grouping of companies within the ITT world-wide system, and having complete control of its own growth policies, STC(SA) will now be free to exploit directly any products, techniques or marketing outlets available within the system calculated to further our local interest. By the same token, we will in the future make our own direct contribution to the system through group area and system committees and by representatives on various task forces set up from time to time to deal with common technical, manufacturing and marketing problems and programmes."

As can be seen in Figure 5-1, such group activities and coordination is accomplished through a regional structure covering Africa and the Middle East (AME), which in turn reports to ITT–Europe, headquartered in London. On technical matters, a Technical Department within ITT–Europe sits in Brussels; it reports to New York

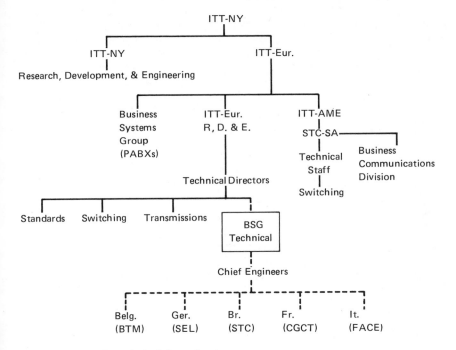

Figure 5-1. ITT Technical Organization.

on all R&D and technical matters. Under ITT–Europe's Technical Department are divisions dealing with standards, switching, transmission, and other telecommunications product lines. The ITTE Business Systems Group has its own Technical Department. It's technical director reports functionally to the director of ITTE's Technical Department. Under the BSG are dotted-line responsibilities for the engineering departments of each of the European companies (Belgium, Britain, France, England, Austria, Italy, and so forth) *and* the BCD in STC–SA. In addition, the technical staff in STC–SA reports on a dotted-line basis to ITTE's Technical Department directly, since ITT–AME has no technical staff.

Because of its small size at present, the BCD of STC–SA reports through the transmissions channels up the line in ITT, rather than directly through the Business Systems Group; as BCD grows, this may change. But on technical matters and decisions as to product development, a different set of channels is employed. The above pattern shows the basic responsibilities, but in practice there is also a "mother-company" for those, such as STC–SA, that are adopting a product developed elsewhere in the ITT system.

The Business Systems Group was formed in 1969 to oversee the

PABXs in Europe and to form a profit center for all private communications along product lines. Therefore, Britain, Spain, Germany, Italy, and Belgium's PABXs are a joint profit center at the level of ITT-Europe (but not Norway or Austria). South Africa joins the conglomerated profits only at the top in ITT-Europe, after going through the transmissions group and AME.

The BSG has a technical director who reports functionally to the head of ITT-Europe's twenty-man technical department. The BSG's Technical Department manages the designs of all PABXs and requires all affiliates to standardize parts so that they are interchangeable. The standardization of piece-parts gives each company the ability to buy from others in the system in emergencies. However, each affiliate has the right to make adaptations locally to meet demands of the national market.

Apart from these lines of communication, there is no departmental structure that provides technical assistance throughout the ITT system; such exchanges are accomplished primarily through personal contact by visits and task forces, formalized by interaffiliate arrangements.

PRODUCT DEVELOPMENT

Prior to 1959, STC only imported PABXs; these were sold largely in the Johannesburg and Pretoria areas. In 1958, the South African Post Office issued specifications for 10- and 20-line PABXs; STC sent these to all ITT affiliates to see whether they were interested. The British affiliate studied the needs of the entire African continent and designed the Stepmaster with 100 lines to meet the high calling rate of 15 to 20 percent trunking (compared to 10 percent in Europe). The Stepmaster was imported for several years under a 30 percent local-content requirement that led to assembly in South Africa. Then the Post Office insisted on a rotary (uni-selector) system, which knocked out the British two-step selector. The Post Office then designed models for 9- and 20- line systems. These were called SAPO I and SAPO II. SEL Germany produced to these designs and shipped the major pieces to STC for assembly according to specifications and direction from SEL; no visits by technicians were required since STC had gained experience assemblying to specifications from Britain. The assembly process involved only the wiring of mounted components. To meet specific requests from the Post Office, two local engineers made modifications.

By 1964, it was clear that STC needed additional models to fill out its line and meet demands for varied capabilities above the SAPO models. The STC engineer was asked to design new PABXs using British standards and pieces; but STC also needed to increase local content, since SAPO had raised this percentage in order to achieve self-sufficiency in communication. STC, therefore, had to reduce the number of assembled items from Germany and began using more British equipment and designs, but they were too expensive. STC turned to the Pentaconta, made by BTM (ITT's affiliate in Belgium) and adapted it to SAPO specifications; 14 kits were imported, but the model was dropped because it had a limited capacity of 3/10/2. The STC technical engineer then took a trip to *all* affiliates in Europe (Portugal, Spain, Belgium, France, Vienna, Italy) and then to ITT-Europe in Brussels to report on his findings. He got all Pentaconta data and determined that it could be adapted further to meet SAPO requirements and customer needs. He then went to London and, under instructions from ITT-Europe, to Spain to order components and to Vienna to order the uni-selectors. At the same time, the engineer negotiated a purchase of all the machinery from the manufacturing departments of various affiliates.

This filled in below the Stepmaster system, capable of 30 to 100 lines. By 1969, the lack of an ITT model with more than 100 lines became critical, since customers wanted one with larger capacity, but SAPO demanded that it *not* use a 2-motion selector as with the Stepmaster. ITT did not have a uni-selector, and STC sought approval of the cross-bar Pentaconta system, which its officials had heard of on the prior trip around ITT. SAPO did not want to approve it, however, for "political" reasons (another company used a cross-bar system and South African policy was then not favorable to that country).

The main developments in PABXs in the ITT system, however, were based on the cross-bar switching system. It appeared more suitable to the South African market, and rotary switching was giving rise to considerable repair problems for the Post Office. The Post Office wanted something different, but it hoped to wait until electronic switching was available (i.e., computer plus cross bar, or simply computer). It was evident in the late 1960s that it would be some years before electronic switching would be available. STC, therefore, set out to persuade the Post Office to accept the cross-bar system, which was more suitable to a system with larger numbers of extension lines, from 100 to 600, and would reduce repairs substantially.

SAPO engineers visited ITT operations in Europe and looked at alternative equipment and its performance. They reportedly "thought highly" of the Pentomat series and were "not prepared to accept any new system that had not been thoroughly proved overseas for some years." The SAPO engineers returned to ask the STC engineers to select the best of the ITT systems. Two STC–SA engineers responsible for switching visited the Italian, Belgian, and Spanish facilities to see what they could readily adopt. One had attended several Task Force meetings in Europe and found that no standard PABX was directly suitable; therefore, no direct help from reports or specifications was available. STC–UK had made a prototype model of a 50-100-line Stantamat, but it was difficult to manufacture; the Spanish Pentomat–40 was attractive; and the Belgium P–1000T was examined. Discussions in Spain showed that the P–250 might be made into modular 50 up to 600; this was the model chosen.

The Pentomat system itself came from an interaffiliate task force called Task Force 9, created in the early 1960s among France, Italy, and Belgium. The basic 50-line module was designed, which could be linked by proper wiring to make a larger system. ITT's Italian affiliate (FACE) never produced the system; BTM of Belgium adopted it, but modified the location of the selector and the shape of the cabinet and took it up to 200 lines. The original design had included modules up to 600 lines. France built a 600-line model, but with different equipment practice, using 24 cubicles rather than 12. Spain finally adopted the original Task Force design, but took it only up to 200 lines. To create its modular system, STC made 16 major modifications and several hundred minor ones.

The STC engineer took the circuitry from the French affiliate and the Spanish equipment practice and designed a means to link multiple Pentomat modules on a "plug-in" basis expanding from a single 50-line module to a 600-line system—a very complex operation and one he was "not supposed to be able to do." His success, he asserted, was a result of his ignorance of the problems of doing it. The more modules to plug together, the more complex the system becomes. He did not report on his efforts thru the usual "R, D, & E Guide" procedure of ITT or get approval to pursue the task; he avows he probably would not have gotten it if he had asked, given the complexity of the task. He had only three men working on it and must have been "mad," he commented, to have attempted it. His efforts were further hampered by the insistence of the marketing department that access to the cubicles be restricted to the front (the French open from both front and back) and by SAPO's insistence that no terminal have more than two connections—if there was a third it had to be a plug-in—to reduce problems in repair. SAPO incorporated a 70 percent local-content requirement in the new systems, and STC–SA could meet this only by obtaining all hardware, wiring, and iron work locally, including high-cost labor. All component piece parts are manufactured in Europe, almost wholly by the Spanish mother-company, CITESA.

The story of the technology transfer necessary to produce the Pentomat system is the major subject of this study of STC (see the following section), but the importance of product development for technology transfer does not end there. STC–SA is seeking new products in the PABX field, and that search requires continuing technology transfer, as will be seen in the final section.

In addition to the 50-line modular Pentomat, STC has been producing a 40-line discrete unit based on the earlier Pentaconta relay and, therefore, involving no new technology. It has also developed an all-relay Pentomat for 5 lines, which can be modularized up to 15 lines. STC engineers studied BTM's 5-line relay system, but decided not to use it; they were able to design their own to meet local requirements. STC abandoned an attempt to develop a lower-cost 100-line unit (not requiring the complexity of plugging two 50-line modules together and removing the cost of two cabinets) because few customers remain at the 100-line level (STC itself has gone from a 50-line PABX to 300-line for its own use) and the cost was not cut much.

Before it decided on the Pentomat for 5 lines, STC asked BTM if they had expanded the P–10; they had—from 3/10/2 to 3/12/2— but it was not enough for STC. They decided they wanted to take the P–10 up to 4/10/4, then down to 2/5/2, and up to a 6/15/6; so they settled on another modular system. They are now in the prototype stage, and other affiliates are likely to adopt the modular system when proven out. However, a gap remained at the 25-line level, so STC looked at BTM's and CITESA's equipment but rejected both. It has designed its own 25-line, cross-bar system, using 30 percent of the equipment from these two affiliates and 70 percent of their own design. Despite their own contributions, STC engineers recognize their dependence on engineering in other affiliates, without which they could not make their own.

Other affiliates, of course, are developing new equipment. The STC engineer was sent to France to look at the new Minimat, with its small cross-bar switch, and to Australia to see electronic controlled switching which was in the prototype stage. STC made no move in either direction, however, especially since ITT will be moving to standardize all electronic equipment over the entire company. But some new design will be desirable, for the present Pentomat-50 has only 6 local links, and new designs provide greater flexibility at that end of the system.

Technology transfers in product development do not, of course, flow just one way—to STC. Its development of the modular 600-line system provides an example of reverse flow back to the mother-company in Spain. The idea of the system was transmitted to Spain through a visit to South Africa and Spain of an ITT–NY engineer. This information prompted a visit to South Africa by two Spanish engineers for two to three months, and they returned with drawings, run-out sheets, layouts, and everything they could carry. An equipment engineer came to find out what STC did to cable forming to permit the linkage between the modules and STC's methods of presentation of information so he could understand the data to be taken to Spain. A circuit engineer was sent to determine the method of expanding circuitry to 600 lines and to familiarize himself with the rewiring necessary in each module to permit the linkages (what was thrown away and what added). Spanish engineers have now manufactured their own modules and have made some slight changes, such as moving the plug to the top of the cabinet rather than inside the back.

In turn, Spain has adapted the system to the needs of hotel PABXs, which require monitoring of calls to cut off malicious ones, interroom switching, single-digit dialing, and multi-digit dialing for multi-storied buildings. STC, in turn informed by the ITT–NY engineer of this development, will send an engineer to Spain to see whether it is suitable for South Africa. These developments demonstrate the continuing flow of technology and technical information on designs, manufacturing, and facilities that occurs in ITT's operations—at no specific charge to the recipient. The rate of transfer varies over time. There is a pattern of technology flow on any new development, which rises very rapidly at the start to a peak in the start-up phase, tapers off as the problems are solved and production gets into full swing, and rises again with modifications, or to a new peak with adoption of a new product.

TRANSFER OF MANUFACTURING TECHNOLOGY: PENTOMAT

PABXs are assembled in a step process in which several components are assembled simultaneously in different areas of the plant and then brought together for final assembly. The platin is a frame in which relays and components are wired in the sequences necessary to relay the messages to the cross-bar for switching. Simultaneously, cables are being formed to make the proper connections between the platins and the cross-bar and the inter-module links to compose the entire system. The cross-bar is also being wired while these processes are completed. The cable harness is then attached in the cabinet, followed by the cross-bar and the platins. No two PABXs are alike; each has different requirements as to interconnections, total system capabilities, and the control system. The transfer of the technology occurs in the design of the system, embodied in the piece parts and components, and in testing.

The layout of the plant for PABXs is not so critical as to require significant technical assistance from the mother-company. STC has redesigned the plant layout to match its own space availabilities with the most efficient flow of parts and components. It used the production layout of Spain in the spring assembly, coil winding, and relays, but not in cable forming or testing.

The machinery for assembly and testing is that used by the mother-company (CITESA) in Spain and is ordered from Marconi in Spain, which supplies CITESA. STC has not, therefore, set up its own machine shop to develop specialized machinery; nor is there apparently much communication among affiliates on the design of machinery. STC gets new information when it orders replacements or new machinery and sees what the manufacturer has done; it may then order adaptations for older machinery.

Design

The major technology transfer lies in the design of the system itself. As noted earlier, the Pentomat was designed by a Task Force of engineers from four ITT affiliates, with five engineers working on the circuitry alone. The four companies involved responded in quite different ways to the opportunities to use the new system.

STC obtained all drawings on the 50-200-line systems from Spain and studied them for three months. They then redrew the design to meet the needs of SAPO and their customers. Then (in 1971) the engineer, the quality-control officer, the maintenance officer, three technicians from manufacturing, and three from the service department went to CITESA for a month. They discussed problem areas and showed CITESA a half dozen shortcomings in the system (for which CITESA engineers were very appreciative). The visit was to permit working with an actual system and to be able to instruct officers who would be responsible for the system at STC and SAPO. Three of the officers—service, quality control, and maintenance training officers—stayed another month.

A second group went to Spain almost immediately on the return of the first and also stayed for a month; there were a design engineer who would become responsible for systems application, an equipment engineer, another quality-control engineer, and a service technician. (Despite the difference in languages, there seemed to be no great difficulty of communication among the engineers, for they could relay on drawings, motions, and demonstration.)

STC then proceeded to draw up the basic system and built a prototype in order to get SAPO approval. More design shortcomings were uncovered and were sent back to CITESA for modification of the system. STC then turned to its own major modification of designing a "plug-in" capability.

The "plug-in" design required the cutting of all connecting cables between the modules of a fixed system and adding plug connections in duplicate on each cabinet, so as to make connections between each and in relay among all—instead of each being wired directly to each of the others. Any trunk line coming into one module is automatically connected to each of the others, and therefore, all 600 lines.

The addition of each 50-line module to the system requires substantial changes, rising in irregular jumps. The greatest changes are with the addition of the fifth and each after six; the sixth requires the addition of between 24 and 30 plug lines, which is small compared to the changes at other stages. But these additions in the plug system must be compared to redoing the entire system when one moves from a 300-line (6 modules) to a 350-line system under the CITESA design.

The pressure that caused STC to come up with the plug-in design was that it faced a constantly changing demand from its customers. CITESA might sell one of its customers a model with over-capacity (a 200-line model when only 150 lines would be used for a time), but STC could not do so for each customer's demand for lines kept rising steadily. A customer might jump from 100-line system to 400-lines in five to ten years; to redo the system completely to expand capacity would be very costly. A plug-in system provides great flexibility by permitting addition of a module—one at a time— as needed.

Processes

The technologies employed by STC in the Pentomat assembly— apart from those embodied in the parts and components imported— are principally in the wiring, coil winding, assembly, adjustment, and testing stages. The transfer of technology is first to the engineer who draws the circuit diagram in response to a customer's needs; for this, there is often an exchange of information among affiliates. For example, STC has received drawings from both BTM and CITESA on Pentomat circuits, which STC had adapted to its own use. The circuit drawings are then clarified by a draftsman for use by foremen and supervisors in instructing operators. An engineer checks out the parts supplies to determine whether the system can be manufactured and another positions the parts on the platin and sets out the interconnection information for the foreman to direct the wiring.

The proper parts are then brought out of stocks for the spring-set assembly, for coil winding (if sufficient coils are not already in stock from prior work schedules); the relays are assembled with the armature; and the set is adjusted, which requires setting of the armature, spacing of contacts, setting spring tension, and checking to see if it operates within the proper limits of current-flow. Simultaneously, components are assembled for joining the relays in the platin, prior to wiring.

Each of these operations requires a significant skill—not only in following instructions but in identifying the right pieces or connections and carrying out the proper operation. A similar skill is required in wiring the cross-bar, which is imported. And, finally, the wiring of the multiswitch, the platins and the inter-cubicle links.

In order to raise cable-forming· productivity, engineers visited ITT facilities everywhere they could to note the procedures; essentially they were formed on nails stuck into a flat board, so that the harness was flat when finished. Wooden hammers were then used to pound it into the shape needed to get it into the cubicle. Spain used a 3-dimensional shape to reduce the flatness of the harness. STC got the idea to force the cables around a three-sided shape *similar* in dimensions to the cubicle into which the harness was to fit; this cut the mounting time by 50 to 60 percent. It is a significant improvement, and STC has informed other affiliates, but does not know whether they have adopted the procedure. STC engineers were also able to alter the platin layouts within the cubicle, thereby getting more into the frame than previously.

Productivity in the platin wiring was increased by the adoption of a "stage-wiring" process in which different colored wires are used for a given number of pages of instructions. The operators do not have to wire the entire platin, only about one-third, dressing their part prior to finally dressing of the entire platin. This repetition in a well-defined job has enabled STC to up-grade the work of the blacks, who have had little education and are quite at home in repetitive tasks.

Standards

None of the changes made in the Pentomat series by STC eliminated the standardization of parts and components within the system.

Rather, ITT's insistence on a high degree of standardization makes it possible for STC to rely on the quality of piece parts and components even if imported from different affiliates. Also, any new development by one affiliate can be readily incorporated by another. Though the requirement to use ITT specifications and standards means that pieces, components, and equipment is reliable, it does not mean that all parts are identical. Standardization is not the same as commonality. Italian parts will not necessarily fit with Belgian parts on STC equipment. But despite the differences in pieces, the PCC effort is worthwhile, for it cuts costs of development considerably, maintains reliability, and prevents mistakes when local manufacture is undertaken.

Not only are parts standardized, but drawings are standardized so that affiliates can understand and customers are given similar instructions. The drawings on the PABXs are quite complex, and SAPO wanted them in great detail and in a form which met its own standards. ITT has been pressing, however, for standardization among its own affiliates to facilitate transfer of information among them— basing its procedures on the recommendations of the International Electrotechnical Commission. It is now hoping that SAPO will go along with these standards.

STC is itself contributing to ITT deliberations on drawings standards, for which monthly meetings have been held, despite the fact that it cannot attend regularly because of the distance. Its role is important simply because it has more experience in microfilming of the drawings (and did attend the meeting on this phase) as a result of SAPO's demand that they be provided in this form. STC is, therefore, providing some reverse technology flow to the other affiliates on this process.

STC is also following up on chance information to cut its cost of drawings. On a visit to a U.K. supplier to the British ITT affiliate, an STC engineer was told that STC–UK (at Harlow) had an excellent system of computer drawing but that none of the other ITT affiliates were using it. The technical director sent engineers to Harlow to see whether STC could send its digital information to Harlow, thereby cutting its costs of design drawing by several thousand Rand.

Testing

Significant skill is also required in the testing procedures that follow ITT instructions laid out by quality-control personnel, though STC uses manual methods instead of electronic or computer testing, simply because the volume does not warrant more sophisticated techniques. The testing procedures are employed all along the line, and once a given series is in production, an assembled platin is taken off the line and sent to engineering for testing.

Quality-control personnel are sent the copies of papers presented to Task Force meetings each year on testing techniques and standard times for testing; but STC is so far away that it seldom attends such meetings. The basic test specifications are laid down by CITESA and by the Piece-parts Control Center (PCC) in France. These procedures are exacting, for *all* lines must be tested.

The sequence of testing is also important, for only sub-assemblies are tested at the plant. The platin, cables, switches, and the cubicles with frame cables are tested separately and sent separately to *central stores*, which has the responsibility of taking them to the customer and assemblying them into the final package. At this point, and only then, can the entire system be tested. There are problems with this procedure, for errors are found and must be corrected on the spot.

The problems arising in the final test are a function of the fact that STC lacks adequate "routiners," which are test procedures based on electronic programs; its volume of production is too low to warrant them. So, instead of adopting an electronic testor developed by CGCT (ITT's French affiliate), STC attempted to design its own, but it could not proceed for lack of adequate engineering staff. The test time at STC, therefore, is still longer than at other affiliates. It has tried to cut it through use of computer programs on a shared-time basis, but one such attempt failed to produce satisfactory results; the programming was simply too complex.

When there is a problem in quality control, STC can rely on assistance from one of several affiliates. CGCT in France is the center for the Pentaconta series; CITESA is for the Pentomat, but it also can go to FACE in Italy or the British affiliate. STC claims to have had little communication directly with ITT–Europe on quality control, for their volume and problems are not significant enough; consequently they do not get the new developments in testing. But they *can* obtain anything they wish, if they ask for it—though some affiliates are more forthcoming than others.

In order to keep abreast, STC engineers have travelled to other affiliates to see test equipment and discuss procedures. One trip covered Spain (CITESA, Marconi, and SESA), France (both the Piece-parts Control Center and LMT), BTM in Belgium, and STK in Oslo. They were seeking the solution to a problem on relay adjustment so as to obtain the best method of reducing costs of adusting and testing. They found it and incidentally picked up significant information on other aspects of quality control; for example, some specifications on equipment and procedures that they did not know existed, information on a semi-automatic adjusting machine, and new test machinery being made at Marconi.

Trips by STC officials outside of the Business Communications Division often turn up something of interest; for example, some new relay-testing equipment. In this instance, they did not adopt the equipment, for a new member of the staff arrived from the Irish affiliate, where a study had been made of the equipment, and it was found to be sorely wanting; adoption would have been a serious mistake for STC. This event illustrates the importance of people-movement in the system. In another case, a new staff member brought information about drilling machines, which caused the manufacturing manager to take a trip to the United States; he learned much there that he had not known even to ask about before and passed this information to a European affiliate on his way back.

Materials

The agreement with CITESA provided for purchase from it of nearly 99 percent of materials used directly in the manufacture of the relays, platins, and components; the cross-bar switching is also imported, already assembled. Local suppliers are used for the wiring, cabinets, frames, and some relay parts. The cross-bar could be purchased from other ITT affiliates, but the Spanish part is cheaper because of a subsidy to exports given by the Spanish government. However, STC has sought information on other sources of supply of piece-parts because it got caught by a strike in Spain on one occasion. In emergencies, STC relies on "Operation Lightening," under which shortages are reported to Brussels every Friday; these are filled from anyone in the family that has an overage. For example, STC obtained spring blades through this channel on several occasions, thereby decreasing its lead time in production. Alternative sources of supply within the ITT system are important to STC over the longer run also, for when it wanted to expand production on one occasion, CITESA could not deliver. It turned to the PCC network to find other sources; PCC swept the alternative suppliers for a variety of piece-parts so as to permit STC to go ahead. Both the standardization of parts and the PCC system are necessary to provide STC flexibility in production.

STC's reliance on imports of parts for PABXs is shown by the sourcing of the items on the following page. In each instance where the piece part or component is imported, the cost of production in South Africa would simply be too high because of the relatively low volume of production.

	Imports (%)	Local (%)
Relay resistors	90	10
Relay capacitors	90[c]	10
Relay diodes	100	
Relay transisters	100	
Relay transformers	50	50
Relay insulators		100
Relay wire		100
Platin frame		100
Cabinet		100
Switch	100	

The comonality of parts is frustrated somewhat by each affiliate making its own modifications for cost reduction and not obtaining PCC approval. For example, Spain changed the shape of the spring blade but sent South Africa both the old and new ones, which of course did not fit together. The U.K. affiliate was selling four different blades in Europe under a single code number in the "conformity book." The extent to which a given affiliate *follows* the standards is, reportedly, not checked by ITT–Europe; rather, other affiliates are relied upon to put pressure on the miscreant when its parts are not interchangeable.

A shift in supply of materials produced new information for STC as to what others were trying out, simply by receipt of the shipment. But, if STC requests to be kept continuingly informed of what others are experimenting with, they get no information. Each affiliate in Europe seeks to get ahead of the others and is reluctant to let others in on new developments. But visits to supplier companies of European affiliates will generally produce information as to what the other ITT companies are doing. For example, the manufacturing manager visited the supplier of machines to an ITT affiliate and picked up information about the materials being fabricated by the machine; he was given samples and the knowledge of how to use the new material. In another instance, he was permitted to visit a Plessey operation in Britain at the request of a machine manufacturer; he then bought the same machine Plessey was using which drills printed circuit boards from numerical tapes.

The interplay of suppliers of materials and machines, therefore, provides STC with information on new developments, but it would not be available but for the family ties with ITT, even though there does not appear to be a free flow information apart from personal visits.

[c]Imported by local suppliers.

Personal Visits

Within ITT, personal visits are the primary mode of exchange of technology. Information flows among companies and through official channels, but it is frequently written up much later than it is put into practice and is considered "old" by those who receive it. STC considers that such information gets to it much too early or too late; it estimates that it adopts only about 10 percent of developments reported in written form through official channels; for example, the "R, D, & E Guides."

The most effective exchange of information occurs, according to officers, at the initiative *of engineering personnel*, who are seeking a particular solution or new development and who have a motive of cost-cutting. STC can call on CITESA for as much information as it wishes, and it will be served, but it is *served best by a personal visit.*

If STC has a "tailored" demand from a customer that it cannot service or provide, it would call or Telex BTM or CITESA to see whether such a system had been provided before, and if so, how. STC would probably then visit the affiliate to see for itself. The sending of drawings are not fully reflective of what takes place in manufacturing processes or procedures. They, therefore, send their engineers to see for themselves and to become acquainted with their opposites; they can then follow up by phone or Telex with a feeling of mutual understanding.

STC officials consider that this cooperation with CITESA is *not* a result of the fact that they buy most of their piece parts from it. They insist that the manufacturing department would be just as cooperative if STC bought nothing, since this department does not consider itself a profit center and is not worried about spending time to help another affiliate. But this cooperation would not exist without a face-to-face visit; not even satellite communication and video will substitute in the view of ITT officials.

The visits of STC engineers to ITT laboratories around the world give them information as to what's going on within ITT worldwide *and* about its competitors, which they could not find out otherwise. On occasion, ITT laboratory experts have been sent from Spain to explain electronic-switching, the networks, and equipment needed for different requirements, though nothing was paid for this service. This is one evidence of the cooperation that exists outside of the pursuit of profit by each affiliate. Such coordination is faciliated by the fact that all European laboratories report directly to the technical director in ITT–Europe and that managing directors must show a cooperative spirit in order to be able to live with each other over the long run.

ITT's task forces on product developments also provide an important opportunity for personal exchange of information. They occur twice a year, and one engineer from STC will attend; he becomes acquainted with recent developments, which are reported in papers presented to the group. This "early warning system" prevents duplication of developmental effort and helps to transmit technical requirements from one affiliate to another. At such meetings, engineering specifications are made available on a new development, and some months later, if the development is to be adopted, quite detailed "international functional specification" (IFS) are sent; but this is still before production takes place. The IFS, which are sent to each affiliate, produce a round-robin discussion.

Assessments of the task force meetings vary from "very useful" to "not so useful," depending on who is reporting and the type of meeting. All indicate that such meetings are the necessary mechanism of establishing personal contacts, so that phone calls and Telexes become effective. But each considers that the other affiliates are "holding something up their sleeve" to maintain a competitive edge over the others; STC may obtain more information because it is small, distant, and not considered a threat. Despite the presentation of technical papers, STC engineers consider that the others do not "bare all"; the engineers reporting on developments are under pressure to "stand out," but at the same time not to tip their hand on "vital developments" they are undertaking. Consequently, the IFS do not reveal all, either.

In addition to product task force meetings, there is a yearly meeting of technical directors, which STC attends, chaired by the ITTE technical director.

There is also a vast amount of written communication about R, D, & E information, 90 percent of which is not relevant to STCs operations in PABXs. Of the 10 percent that is related, STC probably picks up 5 percent. One such idea was a dial tone detector that cut off undesired outgoing calls, developed by the U.K. affiliate; it would be used to prevent long-distance calls being made from unauthorized extensions.

If STC were adopting new products or procedures, ITT-Europe would likely be informed and could decide to send an engineer down with recommendations on new processes or procedures. Since these recommendations would come from wide ITT experience, the STC engineering staff would have to have very good arguments to go against them. Since the STC technical director reports on a "dotted-line" basis of ITT-Europe's R, D & E head, he would not introduce any significant change without checking to see whether it fit into ITTs overall technical direction. Despite the possibility of such visits from ITT-Europe, no one has visited STC from the regional office as yet on PABXs, though there have been visits from an official in Technical Direction at ITT-NY. He has been the channel of transfer of some technology from the United States to STC and among affiliates on PABXs, since he has visited PABX installations on a round-robin basis. He has been able to fill the gap of knowledge on *who* to talk to in other affiliates—that is, who is doing what and who are the best designers or technicians in a particular area. STC can then direct its inquiries more effectively—a critical contribution in easing and expanding technology transfers.

A major problem with such personal visits is the absence of knowledge of precisely which affiliate to visit—where is the development in which STC is interested? Visits tend to become fishing expeditions, with a hope that something interesting will turn up and it most always does! But STC gets no *continuing* support from other affiliates that take an interest in being certain that it gets information on PABX developments; it gets only what it asks for and must rely on clues from technical reports that are circulated among all affiliates. This is not to downgrade the distribution of technical reports, but merely to state that the nature of such transmittal is so general that the exchange of precise information still requires a personal visit. The one affiliate that may have done the most might not have reported fully on its progress in such reports.

Visits by other officials also often produce new information that does not get into the formal pipelines. For example, one marketing official of BCD in STC was visiting the Belgian operation looking for new product ideas. He noted to them that STC was using a different method of cable forming that doubled output in half the space; Belgium immediately picked it up.

If such personal visits are not made, much information is simply not put into the pipeline and would be missed. For example, STC-SA made 16 major modifications in the Pentomat system to remove bugs and several hundred minor ones to improve operation. Its engineers showed these changes to engineers in Belgium, France, and Spain. The Belgian engineers had found some of the problems and employed similar solutions, but had told no one—"no reason to tell the competing ITT French affiliate." The French had found some, but also were not interested in telling the Belgians. Only the Spanish were really interested in what STC had found and were pleased to have the assistance. The Belgians and French were, reportedly, surprised to see that any new idea might come out of the South African affiliate, which was seen as small and "relatively backward" technologically. During such visits, the French engineers would point out weaknesses in the German affiliate's system, warning the STC engineer of them and pointing out specific questions to ask on his visit there; the Germans would do the same for the Belgian equipment, all of which lead to some embarrassing discussions at the other affiliate.

Technology seems to be more readily transferred among affiliates who are not in competition with each other. STC reports that the European affiliates are generally not as communicative among themselves as they seem to be with STC, who they do not see as any threat. STC has always obtained full cooperation from manufacturing or engineering staffs in any European affiliate, but this is because they use personal contacts made at the various meetings. Their personal visits have almost always produced considerable new and useful information for STC, but they have also provided some "odd disappointments," when information desired was simply not available or was "talked around" by the host engineers.

The exchange in such visits are not always wholly open and frank, depending on the composition of the meeting. STC engineers must be cautious in judging what they hear; for example, if they are seeking information on a new development they might want to introduce, they are frequently told that it would be too costly for them to institute it on their own ("fantastically expensive") but that the European affiliate could readily supply equipment or components for this purpose to STC. STC engineers report that it is difficult to get a completely unbiased opinion from any European affiliate they have visited; each is seeking its own interest, especially when marketing departments are involved. It takes some sophistication on their part to put the information in perspective. It is their view that their smaller operation can be more imaginative, since less specialized, and therefore less myopic about its alternatives.

Visits abroad by STC engineers since 1960 on PABXs have amounted to eight trips by the technical staff and three or four by the chief engineer; in addition, another four were made on switching problems. During none of these could the STC engineers convince the French, Belgians, or British that they could learn anything by a visit to South Africa; however, the Spanish did send two engineers down on the extension of Pentomat to 600 lines. Complementing the STC visits were a few from the European affiliates to Johannesburg to assist STC: one design engineer came from the United Kingdom on the Stepmaster (the first of the line) in early 1960s and one from the New York technical director.

Nor would assistance from another affiliate in the form of specifications on manufacturing processes, for instance, be adequate; shop practice does not follow the specifications since modifications are introduced on the floor that do not get into the drawings or procedure manuals. Only personal visits will uncover these "little-black-book" procedures. There is, therefore, no standard "manufacturing practice" manual for all ITT affiliates; all are different. The "little-black-book" tactic extends even within a company. In a visit to the French company, STC found that one modification it had made on double-switching under heavy traffic had been discovered and used by the installer-engineers, but they had not told the production engineers in the same company.

Sometimes information comes by circuitous routes. In one instance, a British trade mission to South Africa included a company making plastic insulators for relays; since it was selling them to the ITT Scottish affiliate, it offered them to STC–SA. STC decided they would be good and asked ITT's Piece-parts Control Center for approval; PCC did not know that the Scottish affiliate was using them. In another, a new staff man from Northern Ireland's affiliate brought substantial information on new manufacturing processes used there; he brought a new terminal-wire wrapping tool, redesigned the work station for bar wiring and the layout of flow lines, and set new standards on the Pentomat. Much information, therefore, is by the back-door or incidental to other purposes, but is nonetheless important and could not be obtained without the family ties that gave rise to the contacts.

Language is a special barrier to technology transfer—a "ringer" is not always a "bell" (it's a relay)—especially at lower engineering levels when written exchanges are used. STC's engineers can work best with Belgians, for they can understand English and Africans. Of some help has been an international dictionary of terms in English translation, prepared by an engineer in ITT–NY, but not all useful ones are yet standardized.

TRAINING

The transfer of technology through training at STC benefits from the extensive training programs and manuals provided by ITT to all affiliates. The programs and promotions resulting from them are different at the various levels of responsibility within the division.

Workers

The requirements for new workers (or replacements) are sent by each department to the personnel section on "recruitment and training." This section tests applicants (of which there are always some standing at the gate or on file) for dexterity, eyesight, memory, precision, health, and intelligence. STC would like applicants to have seven or eight years of school (or Standard 6), but the tests could be passed by someone with five years if he can read and write well enough.

To up-grade workers, STC has joined other companies in supporting an institute run by the Catholic Church for Africans, Asians, and Coloreds and in providing adult schooling at night to over 1,000 persons. It has courses on electronics, basic electricity, and mechanics; Africans travel 20 miles to attend. The company gives R800 per year to anyone taking courses while employed; only one African and five Europeans were taking advantage of this offer in 1974 to pursue job-related courses, in subjects as diverse as social science and engineering. In addition, children of black employees are assisted in completing the last two years of high school, which becomes very costly to the African. The company pays for tuition, board, and books; but few take advantage of it because they do not want to stay in school when job opportunities are already good at that level. (The company also gives $2,000 per year to the Institute on Race Relations to help employees of the company and others.)

STC has employees of all four backgrounds in the country; black, colored, Asian, and European. Blacks have been equally productive once they acquire the skills as the others, though they are a bit slower in picking them up.

The training program given each new employee in wiring begins with some elementary theory as to what the platins do, followed by exercises in counting off the tag-blocks from a diagram so that they will fit the wires onto the right terminals. This requires reading and counting skills at fairly rudimentary levels, but the diagrams must be read accurately. They are then taught how to skin a wire to expose the contact area, how to use the power tools to feed in the wire and wrap it around the terminal (tight enough but not too tight), and how to dress the wire (wrapping the many wires into neat bundles). After training, practice is provided on the line, but under supervision, to get up to productivity levels; when a worker reaches 80 percent of the shop level, he goes off of the training program.

Different training programs are given for work on relay assembly and adjustment. Workers will be taught the winding of the armature bars, spring assembly, and winding. Relay adjustment, which is taught on the shop floor, requires two days on trials to get experience; it requires a very delicate touch in determining the tension of the springs. With practice the worker develops precision and speed, but it requires three months on the floor to get up to shop productivity levels. Most of the women are quite capable and can make all the necessary adjustments in *one* operation.

It requires about a week to learn spring assembly—that is, identifying the right spring in sequence and then putting them together properly—and speed comes only with practice. Coil winding is taught in the shop; it requires learning how to fit wire onto spools, choosing the right guage wire, and maintaining the right tension while winding; machines do the actual winding but they must be set properly, according to information given the worker. This requires no more than one day in instruction.

Workers are trained for a particular task and are seldom shifted to other assignments; though a worker may be shifted if he or she requests it, or if there is a surplus of workers in one operation and they need to be moved to others to even out production. They will then need retraining. Productivity is higher than in Spain simply because of low turnover in jobs.

Workers in cable forming are men, for it is heavy and physical as well as mentally exacting. At present they are trained on the floor, for there are not enough replacements or enough new men to require a classroom approach. This task was formerly done by Europeans only but is now done also by Asians and will soon be done by blacks also. It requires about three months to learn.

The cable forming process is divided into stages so that each worker gets about twenty pages of instructions twice daily, with each taking about four hours to complete. He can determine his own pace—working slowly, taking rests, or quitting early, or getting ahead so as to have time off later if he wishes—so long as he completes his two twenty-page mock-up. (Blacks reportedly will not overfill production schedules even if given a bonus to do so; they have no concept of planning ahead.)

Wiring of the cross-bar is similar to that of the platin, but it requires greater expertise, for there are more wires to be attached. A good worker here can move up into cubicle wiring, where the reading requirements are still greater, for 6,000 different wires must be fit to terminals once all the platins, components, and cross-bar are in place. To reduce the complexity, this process is divided into stages, with each worker having only about 1,000 wires to attach. The process also has to be tested in stages.

Promotion of workers occurs within the same work section they have been trained. A competent worker and one capable of checking out the work of others will be named a section leader, with ten people under him or her. At present, because of the recent hiring of blacks in production, Asian or colored women tend to be section leaders; but this will change, for each group prefers to have members of its own group in charge—Asians over Asians, coloreds over coloreds, and blacks over blacks. This breaks down at the next level of supervision, which tends to be composed of Europeans.

Supervisors and Foremen

At present supervisors and foremen are almost wholly European, with a supervisor having two section leaders under her or him, and a foreman being in charge of several supervisors (depending on the operation). It is expected that there will soon be women (Asian) foremen over women (Asian) workers in both the coil winding and wiring operations, but that men will remain in cable forming with men as foremen.

Foremen generally have at least ten years of experience and need a Standard Junior Certificate; they must be able to interpret all engineering instructions. A supervisor should have a tenth-grade (Standard 8) education and a minimum of three years experience; he only needs to know the specific operation over which he has charge.

Engineers

Engineers come to STC with at least the National Technical Certificate, which means that they are really technicians and not graduate engineers; some few have university degrees. The Technical Certificate means that the engineers have been in classes for three years, and some for five. Their courses include topics such as mathematics, technical electricity, telephone and allied communications subjects. Once hired, they get an intensive in-house course on switching circuiting and facilities of the PABXs.

The chief engineer and one or two of his engineers have been to Europe on more than one occasion to visit affiliates there. At the staff level in STC, there are two other engineers responsible for developments in PABXs. Within BCD there is a chief engineer who has charge of three design engineers, two equipment engineers, two customer applications engineers, plus one draftsman, and one tracer of drawings.

South Africa is not well supplied with people who can handle the expanding technical problems of industrialization, and once trained (or if he can merely acquire experience, without training) an engineer or technician can get a job most anywhere, at higher salary. His experience need not have any particular relevance to the new task. Consequently, STC takes technicians for engineering positions, places men in foreman positions who have never been in a factory before, promotes personnel before they are really prepared, and is constantly recruiting overseas for draftsmen, engineers, and so forth. The result of these practices is high turnover because of other opportunities or of dissatisfaction in a position that is complex for the individual (the Peter Principle operates quickly in South Africa).

This pressure on skilled workers, technicians, and engineers is a result of both the population structure of the country and its delayed industrialization, for it was held back by colonial economic policies of Britain and is just now pushing ahead—rapidly! There is increasing pressure, therefore, to substitute blacks for coloreds, thereby pushing the latter up the scale, who in turn push up the Asians and Europeans. But the pushing is faster than the preparation. At the top levels, engineers are now moving into commercial courses at universities (night schools, and so forth) in order to undertake sales and managerial positions. (As with many countries in the early industrial stages, sales is where the money is.) This constitutes yet another drain on engineering manpower. To help top-level personnel advance within the company, STC has instituted an Assessment Center for all grade 16s and above (the top 100 out of 2,600 employees in the entire company) since it is feasible for engineers and managers in BCD to move into transmission, and vice versa—as has occurred. In addition, some STC managers have attended the ITT training managers conferences, given every two years to sharpen training techniques; the subjects include management training courses, engineering and sales training, plus review of the training manual.

The significance of these comments is that, without the ability to tie into the large engineering reservoir within ITT, STC simply could not progress technically and provide new employment opportunities.

DISSEMINATION OF TECHNOLOGY

The dissemination of technology and technical skills throughout the economy occurs through training of SAPO technicians, assistance to suppliers, turnover of workers, and changes in the lifestyles of workers who incorporate some of their learning into their everyday life.

SAPO Technicians
SAPO maintains and repairs all of the PABXs purchased by private customers. It is, therefore, necessary to train them in the

cross-bar system (Pentomat), for there is no other supplier of this system, and it is not possible to obtain know-how on specific switching systems in the technical institutes since each system is tailored to customer needs. On his second trip to Europe, an STC engineer sought assistance on preparation of a handbook for SAPO technicians; his rewritten version concentrated on "faults," and it is now used throughout ITT.

The director of the training section in STC (who was among the first group going to CITESA) goes out to the different Post Offices around the country to train technicians prior to the installation of the equipment by STC engineers. (Once installed, SAPO takes over.) This training is now solely on the cross-bar system, but five and six years ago STC had to supply the same training on relay adjustments and the use of adjustment tools; this required a month of travel.

Training on the cross-bar is done through flip charts and diagrams, plus a small-scale switch model with a complete switch. A full day is required to explain the contacts and switching mechanism, then four more days on circuitry and the facilities available on the P–40 and P–600. (Such training, of course, facilitates acceptance of the next model, the P–25 modular system.)

The SAPO technicians that receive this training are those doing the actual repairs, plus the Control Technician. In Pretoria, the course was attended by the chief technician and the senior technician as well as seven repair technicians; even so the supervisors would not be able to teach the next group of their own technicians. The STC instructor has trained over 100 technicians over the past three years throughout South Africa: 12 in the Rand, 9 in Pretoria, 6 in the Transvaal, 14 at Bloemfontain, 5 at Kimberly (including a de Beers mine technician, for he maintains that system), 25 at Capetown, 7 at East London, 6 at Durban, 5 at PieterMaritzburg, 7 in the Central Matal area, and 10 at Port Elizabeth.

Suppliers

The more sophisticated materials are, of course, imported; only five suppliers are local and provide relay winding wire, cable receiving, and cabinet mechanical parts. Some suppliers do need assistance. The cabinetmaker literally produces to the Spanish drawings (retraced and translated by STC). The power supply for the Pentomat was designed by STC for manufacture by a local supplier, and the manufacturing department has helped a local machine shop diversify its operations. It was making "tailored" machinery, and STC engineers showed it how it could make piece parts ready for assembly as needed by customers, such as themselves. Consequently, it is now serving more customers and is regularizing employment. STC is also moving into plastic piece-parts and can get some of them locally. It has given all information to the supplier to make test samples; STC is not capable of telling a supplier *how* to make them or to do better, but it can put him in touch with European suppliers (such as ICI) who serve European affiliates. In one case, the European supplier came down to South Africa on request; it explained that ITT was using a new and different material and helped the local company move into that material.

Turnover

In some companies, the turnover of workers provide a dissemination of skills, but STC has found that the insecurity among blacks and coloreds reduces their turnover substantially. This is in sharp contrast to the 40 percent turnover they have had in the past with a larger percentage of European workers; with the loss of Europeans, they have hired more blacks and coloreds. Their entrance standards have dropped from a hoped-for Standard 8 (10 years of schooling) down as far as Standard 2 (4 years); though they still search for Standard 4 and Standard 6, they cannot expect to find them.

STC has adopted a Turnover and Absentee Program (TAP) to seek to reduce these occurrences; it concentrates on problems faced by workers and seems to be of some help. Despite the gentlemen's agreement against raiding, engineers and technicians do go to the competition—often through a third company to "wash" their association with STC.

To help reduce turnover of supervisors by helping them adjust to the problems of an industrial company, overtime, and so forth, the supervisors and foremen get a "management program" that focuses on the impacts of managerial responsibilities on home life, changing schedules, travel and so forth.

Lifestyles

Technology can be disseminated also through the everyday life of the workers of the company, in which they practice some of the skills learned or the orientations to life developed at work. Several of the workers, including a member of the Joint Worker–Management Consultative Committee (who had recently polled many of the workers on his own initiative to learn their attitudes), cited some examples of this dissemination. There is nothing sophisticated about this skill-transmission, but it is more than would have occured otherwise and is a contribution to improvement of lifestyles.

One Indian woman had four children (husband and parents in the family, with the husband working): two children were in high school, with company support, and will complete Standard 10. She has been in STC only one year and has worked her way up from sign painting, to coils, to bar-wiring, to platin wiring, and section leader. She has taken a strong interest in helping others learn skills, starting with reading, and is helping teach children to read in her school community. At an elementary level, she has been able to repair simple appliances from learning about wiring techniques at STC.

An Indian supervisor of cubicle wiring rose from harness assembly in three years; he has 30 employees under him, of which 23 are women and require "careful handling," which he is proud to report has produced the lowest turnover in any area of BCD. He keeps his company card on "how to handle people" in his wallet and refers to it in any conflict or negotiating situation (it is actually well worn). He uses these techniques in handling football players and drunks on a weekend. They have been particularly useful in his relations with his wife, producing a happier home life, especially when he has to work overtime and misses dinner or some planned occasion. He also has employed his technical skills in appliance repair and minor auto repair in the neighborhood; this ability has given him confidence and an eagerness to observe other technical operations so as to learn more.

The workers themselves explained the lower turnover among blacks, coloreds, and Asians as a result of good working conditions and much better employment opportunities and treatment than they have had in the past. Their friends, reportedly, cannot understand how one could find employment in a company that subsidizes tour vacations to Swaziland and Botswana, that provides free medical care, gives free meals and transportation if there is overtime work, that covers all employees by a pension, where all elements of the population can work in harmony under a single roof, where the cafeteria offers a variety of meals including both European and African style food, where classes in dancing and other social graces are offered to all after work, that provides family-planning instruction, that pays employees while in training, that supports children in school, and so forth. Each of these contributions gives the workers a stronger desire to improve themselves and continue with training. But the proof of the effects is not yet in for the blacks, since they have only recently been brought into the production force and have not yet had time to respond fully to these inducements.

GROWTH POTENTIAL

As evidenced by STC's continued product development in PABXs, one of the more fundamental contributions of technology transfer is growth through new products, increasing employment opportunities and meeting new demands. BCD has, therefore, been examining the expansion and changes in the market in South Africa (and Europe) and investigating the new product developments within ITT as well as by the competition. For example, Plessey and Siemens are introducing electronic equipment, and all are looking toward the *public* switching systems as a new market in South Africa.

Market

STC considers that 35 to 40 percent of the market for PABXs is about all that it can hope to obtain (compared to its present 25--30 percent) given the present competitive structure; to gain more it would have to engage in strong price cutting (which it cannot afford) and cause considerable tension among competitors; this is not the South African way. It will need both new PABXs and new products to continue its growth. The Pentomat will be adequate for *its* job for a couple of decades but will be outmoded eventually by new technology. BCD, therefore, sought guidance from ITT-AME and ITT-Europe as to whom to see on an exploratory trip in mid-1974. It received excellent guidance, and its officers were given a thorough briefing as to new product developments in Europe.

The division is not restricting itself to PABXs or even to its present customers; it has recognized that it can call on the vast production experience of ITT from which to chose and could, therefore, go into public switching equipment for the SAPO or into data systems or peripherals. SAPO has asked for bids on curing its own switching problems, which arise partly from its use of three systems from separate companies (STC is not among them), and SAPO wants a single electronic system. ITT has offered its Metaconta System; Philips, CIT, Plessy, and Siemens have bid. SAPO has insisted on 99 percent local content within 10 years, which means that STC would have to produce or obtain locally what does not now exist in South Africa.

STC could start with 70 percent and go up only with local production of diodes, integrated circuits, memories, and computer terminals, which are produced in a very few countries; adding them would be very expensive. Even relays require imported glass and components, though these could be stockpiled to meet the strategic considerations of the government. (ITT has calculated that 100 percent security of supply can be gained only with an electro-mechanical system, but at four times the world price.) If the ITT bid was accepted, it would have to license one or two other companies (TMSA and Siemens) to produce parts of the system so as to keep them alive. This bid is wholly dependent on technology developed in France, London, and Germany—working under ITT–Europe direction. STC would not even be in the bidding without ITT backing, and ITT would not be interested in licensing an independent company to produce the system. STC's mother-company would be the French LMT, which would back-stop it completely. Even the prebid negotiations required both French and Belgian technicians to visit STC for weeks.

Another development opens an alternative route for growth— pulse code modulation, which will tend to tie long-line transmission, main switching systems, and PABXs into a single system and force interface equipment to be developed. STC is already in two of the parts and could readily move into the other if the market opened up satisfactorily in South Africa. Again, such a move would be wholly dependent on the technology resources of ITT.

A third alternative is the move into data systems. This would involve bringing some ITT products into South Africa for sale, since they would simply be too costly to produce locally and would be obsolete before full production could be achieved. The overall world market is growing 50 percent per year, but there are many competitors, and a small operation could not survive without protection that would make the system for the customer uneconomic to purchase.

Such a system would require, however, interfaces with PABXs; so an attempt could be made to marry data systems and switching. IBM has an integrated PABX–data system; ITT has separated them with the result of lower cost; Siemens has equipment both ways. The objective, of course, is to transfer data from one point to another in useful form. ITT has all of the components, though they are not all in the same advanced stages of development and production. Much investigation and work is necessary before a decision will be made in this direction. The trip in 1974 was for the purpose of starting such an investigation and to examine other possibilities that might arise during the conversations.

Research Trip

For a month, the director of BCD and his marketing research manager visited ITT affiliates in Europe and the headquarters in New York to acquire their thinking on long-range PABXs and data equipment sales. They started with ITT–Europe's Business Systems Group and continued to the data exhibition at the Hannover Fair; to Stockholm to see Modems and PABXs; to Oslo for Cryptel, PABXs, and CADEM; to ITT—AME (London) on PABX and data projections; and to ITT–NY for similar discussions; back to Britain to discuss data and tele-printers, ADX, and terminals; on to the Hague to discuss data mar-keting; to Stuttgart on the ESR and data terminals; and finally to Milan on PAX and Achirophones.

They returned with a much better appreciation of the possibil-ities of integrated "voice, message, data" systems (requiring PABXs, data collection, transmission, and storage), which can be programmed for any combination of applications. It would appear that involve-ment in data systems will become necessary to maintain STC's posi-tion in switching. The key, of course, will be in software, for there are many hardware alternatives available. STC must plan its needs for staffing to support this new move; it can rely somewhat on the experience of STC–Australia, which went through the same exercise four years ago and is now successfully exploiting this market—a fact STC learned only on the visit to New York.

The trip produced some 30 different follow-up actions on the part of BCD to make the best use of the information and materials obtained on the trip. One of the most important was to consolidate their contacts with key individuals with whom they had talked, requesting more information and sending some that had been re-quested of STC. This alone was a large task, for they talked with some 8 to 12 individuals, ranging from top managers, to engineers, and marketing personnel, at each affiliate.

These resources would simply not have been available to STC if it were not a member of the ITT family. Not only does South Africa lack the resources, but also no independent company would have been given the same reception or as full a disclosure of information as was obtained by the BCD officials on this trip. The mere volume of written information was staggering, covering a five-foot book shelf. And the value of the interpretations given orally cannot be assessed; they would not have been available to any market re-searcher employable by STC independently. For example, at Brussels they discussed with the product-line managers the problems of switching, computer use, remote control, and so forth; they received a full briefing on ITT policy on data systems, on the systems them-selves, on projected market size, on competitors, on technical trends and provided manuscripts on the same topics.

The written materials were composed about one-third by internal reports within ITT on products and markets and about two-thirds of promotional materials explaining the use and marketing of the systems. Even the information contained in published reports sent them later would not have been known to them except for these contacts, for example, a copy of a consultant's work on the future market for computer terminals (the ITT host gave the visitors his only copy and ordered another for himself). The documents obtained from affiliates covered such important topics as the "Data Terminal Marketing Plan, 1974–1976," "Telecommunications Services for Handicapped," the "ITT System 710 Task Force Report," marketing specifications, technical specifications, "Voice-data-software marketing requirements," and a variety of materials on peripherals.

As a result of these visits, STC officials have personal contacts they can rely upon to follow up by letter or Telex and will feel comfortable that their objectives will be understood. They were well received by all affiliates, though in one case a closed entry had to be reopened by a call from ITT--Europe. Such a round-robin visit was absolutely necessary, since there is no Task Force on data systems and PABXs. With the visit, BCD was able to come up with quick recommendations as to next steps and how to proceed, coupling their new information with local research results.

The ability of these officials to visit several affiliates provided a picture that could not have been gotten from a visit to only one mother-company. No one affiliate seemed to have the entire picture; it had to be pieced together at both the marketing and technical level. After these visits, BCD now feels it is desirable to follow ITT–New York's recommendation that they go to Australia to see how a company similar to theirs has penetrated the market; the European companies are too advanced to be a good model for STC. Belonging to the family of ITT affiliates provides the variety of knowledge and experience out of which STC can grow soundly and with adequate technical back-up.

Chapter 6

ITT-Mexico: Industria de Telecomunicaciones S.A. (INDETEL)

The history of INDETEL demonstrates the evolution of a technology cooperation between ITT, the Mexican government, and the local infrastructure that resulted in an extensive new technological capability for Mexico and a successful business growth for ITT. The case involves ITT's transfer of technology to Mexico for production of telephone subsets (home telephone sets) and clearly demonstrates the different cycles of transfer and how a technology base can be developed over time, thereby leading finally to local research and development in a wide variety of products. As a result, INDETEL, the ITT subsidiary that produces subsets, transmission equipment, and switching devices in Mexico, grew from a joint venture distributor of equipment, to a large production and product development center. ITT's expert mix of documentation, exchange of personnel, and intercorporate project collaboration were the cornerstones on which the company built up a steady and ever-increasing flow of technology into Mexico. The strength of the system is demonstrated by the fact that although in 1974 ITT reduced its ownership of INDETEL to a minority position, it continues to be the major supplier of technology to the operation.

HISTORICAL DEVELOPMENT OF TECHNOLOGY BASE

Until 1964, ITT operated in Mexico as a joint venture partner of Erickson and supplied assemblies and components for the National

Telephone System. Most of the transmission, switching, and subscriber equipment was imported by INDETEL from another ITT subsidiary in Belgium. Contractors worked with the Mexican government to install and maintain the different elements of the national telephone system.

To provide basic telephone service, a telecommunications system must contain subsets, switching devices, transmission equipment, and wire and cable. At the heart of the system is the "central office," which provides the complex services of *switching* and routing of the telephone signals between the millions of users and trunk lines. The signals are sent, amplified, and modulated by the *transmission* equipment. At the receiving or sending end of the network, individual *subscriber subsets* (telephones) may be further controlled or rerouted in the office or building by the use of private switching (PBX), secretarial sets, or other equipment that can hold, monitor, or redirect the signals. The millions of subsets, the switching apparatus, and other equipment are linked together by a variety of *wires and cables*, known as the "outside plant." INDETEL's business mission in Mexico has been to produce and develop locally as much of this equipment as possible, provide continuous new technology for system improvement and cost reduction, and deliver extensive maintenance support to repair and replace previously installed products.

Until 1959, INDETEL produced few of the 400 components required for a telephone and confined its industrial activity to assembly of imported components. But by 1959 a few parts and sub-assemblies were manufactured at the plant including brackets, baseplates, and some housing pieces.

This initial production effort stimulated INDETEL to begin minor modifications in the subset. For example, INDETEL designed special holes in the ringer for tone modification and altered the subset housing for the Mexican system. Production of subsets increased steadily, from 10,000 in 1954, 25,000 in 1959, to 270,000 sets in 1974. Additionally, INDETEL repaired more than 230,000 subsets in 1974.

INDETEL's drive to increase local content in production was impeded by lack of local vendors for supplies and contractors for special fabrication. ITT staff worked continuously with them on areas such as quality control, cost effectiveness and production planning to enable the local manufacturers to support the integration effort. With the improvement of local supplies in capacitors, resistors, and other components, the plant was able to take over production of additional subassemblies. By 1960, the click suppressor and new types of plastic moulding could be supplied by Mexican manufacturers.

The growth of the INDETEL plant and its local infrastructure lead to greater involvement in product development. The secretarial set, a new form of electro-mechanical telephone, was tested and evaluated in Mexico at this time. Tel-Mex, the government agency in charge of telephone communication, worked closely with ITT to integrate this new type of subset into the Mexican system. By 1962, both the secretarial set and a new wall set were being produced in Mexico as a result of technology transferred from the ITT plant at Corinth Mississippi, and ITT's Bell Telephone Manufacturing (BTM) in Belgium.

After acquiring 50 percent of INDETEL in 1963 from Erickson, ITT accelerated its programs to increase the amount of local content in the telecommunications equipment it produced in Mexico. ITT cooperated with the government to develop a new set of specifications for the national telephone system and made a push to provide the Mexicans with basic technology for switching systems to complement the production of subsets. The new specifications were to be met by an improved subscriber subset called the "Sonofon," and a new facility was constructed at Toluca, outside of Mexico City, to produce advanced types of switching equipment.

Between 1966 and 1969, INDETEL made still more extensive modifications on the subset, with emphasis on reworking the network and producing new types of housing. Ringers were produced under license from Erickson and Toshiba, and the Mexican plant took over complete production of the hook-switch subassembly. Technology for this component came for Compania Internacional de Telecomunicaciones y Electronica, S.A. (CITESA), the ITT subsidiary in Madrid.

Between 1970 and 1974, the INDETEL staff increased its production and supply efforts until over 90 percent of the Sonofon subset was being produced locally. A local-content level of 65 percent was reached for the private switching devices (PBX), and the products emerging in the areas of advanced switching, channelling and multiplexing were near 55 percent local content. In addition, the Mexican operation could now boast an extensive internal research, development, and engineering capability that was receiving many new research programs from the ITT Global R&D activity. Mexican engineers had completely reworked the components of the Sonofon subset, adapted a new dial, and developed a special design for the Mexican market. Only the gears in the dial were still imported. Value engineering had also led to highly specialized manufacturing procedures unique to the INDETEL plant.

The growth of the INDETEL's technological capability paralleled the market and infrastructural development of Mexico. INDETEL's overall capability to create and utilize technology was reflected in expanding sales and staff, as shown in Table 6-1.

Table 6-1. Indetel Sales and Personnel Growth, 1968-1972

	1968	1969	1970	1971	1972	1973
Personnel	672	863	1224	1840	2473	3863
Sales (millions of 1973 pesos)	71	83	111	158	302	438
Salaries & Benefits (millions of 1973 pesos)	21	29	43	78	115	155

New production facilities were constructed in 1974 in Cautitlan to house the increased manufacturing and expanded research and development activities. The improved products and technology capability evolved at a pace dictated by the market demand for the products, the ability of the infrastructure of Mexico to receive and support technology, and the global capability of ITT to mobilize and direct appropriate technology to INDETEL. The progress of the Mexican subsidiary from its original assembly activities, through increased integration, to product development and research are linked to a gradual accumulation of know-how necessary to command different elements of the subset production. By 1974, INDETEL was looked to by ITT headquarters and the other regional units as the most advanced in Latin America for subset technology.

PRODUCTS AND MANUFACTURING

In 1974, INDETEL produced four major telecommunication product lines at its two plants: central switching, private switching (PBX), subscriber telephone sets, and transmission systems. Central switching products represent a new area of expansion for the company, which hopes to fully implement ITT's Pentaconta semi-electronic systems throughout Mexico. Relative to major public switching equipment, the Pentomat private switching equipment is produced on a small scale. For transmission, new multiplex and line transmission equipment is also produced in the Toluca plant.

At the Cautitlan plant, basic engineering is conducted and the subscriber subsets are manufactured. Sales and service of these subsets represented approximately 25 percent of total sales for 1974. Manufacture of the subscriber subsets occupied approximately

16 percent of total employees in 1974, compared to 28 percent in 1968. The proportion of manpower devoted to the Sonofon development and production had diminished in relation to the total since 1968 as greater emphasis began to be directed to switching equipment. The deployment of INDETEL's personnel in 1968 compared to 1974 and anticipated 1975 is illustrated in Table 6-2.

Table 6-2. Allocation of INDETEL Personnel

	1968	1974	1975
Switching	225	1341	1395
Subsets	194	515	545
Transmission		114	104
Central Office Exchange	236	725	798
Central Manufacturing			210
Engineering		285	304
Administration & Marketing	17	156	229
Total	672	3136	3585

However, the evolution of the Sonofon subset formed the basis for INDETEL'S corporate growth in technology and its capacity to absorb the switching technology introduced in the '70s.

Manufacturing Subscriber Subsets

Most people fail to realize the complexity of services rendered by a telephone subset. Simply speaking, the individual subscriber subset must be able to: (1) transmit, (2) receive, (3) ring or otherwise call attention, (4) monitor transmissions, and, of course, (5) withstand extreme wear (at least 200,000 removals of the hand set, for example). Within subset technology, input from such widely divergent fields as plastics, metalurgy, electronics, and mechanical engineering is involved. The simplest home subscriber subset contains about 400 distinct components.

The Cautitlan plant produces unique dyes and tools and fabricates plates and special parts. Over ten different metals are used for the bells, brackets, minute spiders, base plates, and other brass, copper, zinc, and iron alloy components. A wide variety of skilled machinists, welders, tool and dye makers are involved in the fabrication of the initial parts.

For the variety of plastic pieces and housings a special plastics area is required. Several types of injection moulding machines are used to combine about a dozen varieties of advanced plastic mixes to meet different specifications. The extremely sensitive carbon transmitter capsules and receiver capsules are assembled and tested in closely controlled "clean rooms" by the most highly skilled assemblers and technicians. The plastics mixing and molding tasks also require skilled machine operators, but the ringer assembly and other sub-assembly activities on the production lines call for only moderate- to low-skilled workers. The specialized clean-room techniques that were modified and perfected by Mexico are now being used by several other ITT production groups.

The Mexican Sonofon contains nine distinct sub-assemblies. From the initial fabrication of metal and plastic parts the various production lines assemble individually the ringer, carbon transmitter, dynamic receiver capsule, dial, handset, hookswitch, printed circuit and wiring network, housing, and base. The nine basic sub-assemblies are then assembled starting with the lower housing to which is attached the hookswitch network and base. The ringer is then inserted followed by the attachment of the handset, dial, transmitter, receiver, and various caps. The final stage is the connection of the wire leads and product testing. Though each component and sub-assembly is individually tested, a final test is always run to evaluate the dial, ringer, transmission and receiver capsules. If flawed, the set is returned for reworking and special diagnosis; otherwise, it is put into the plastic housing and boxed for delivery.

TECHNOLOGY AND ITS TRANSFER

During the ten-year technology expansion of INDETEL from 1964 to 1974, a wide variety of technologies and methods of transfer were utilized. As in most cases of industrial technology, the basic technologies come in the form of original design technology, the technology planning and organizing capabilities, production methods and start-up techniques, and the on-going stages of value engineering and product development. In Mexico, the transfer evolved gradually, constrained by the ability of the subsidiary and its infrastructure to absorb and utilize technology. Through the broadening application of technology from documentation, exchanges, training and other sources, INDETEL moved to a position of almost 100 percent local content in subset production and established a major research and product development laboratory serving Mexico and the ITT global system.

Technology Management and Planning

From the inception of the subset assembly in the '50s, ITT brought its control and planning technologies to play. The use of specialized planning experts and systems were used to pinpoint the stages at which local production, product development, and new market services would be introduced. Though this type of support is often less visible, it is crucial to the successful transfer and absorption of technology. Too rapid or imbalanced progress will not only disrupt human resource and system development, but also incur unnecessary costs and failures.

A good example of the importance of planning and control support is demonstrated in plans for the introduction of new types of telephone subsets and electronic switching systems. Both the touch-tone and quick-step-dial phone subsets could have been adapted to the new switching being used in Mexico. The company preferred the more advanced electronic system embodied in the quick-step technology for certain parts of Mexico, even though the price to deliver the ordinary phone was approximately $25 per unit, while the quick-step could run as much as $70. However, the quick-step approach offered many advantages that appealed to the Mexican government, even though there existed such a wide price differential at the individual unit level.

ITT headquarters staff participated in evaluating the technology and business program for the new venture and was able to suggest a way of introducing the advanced technology at a lower initial cost. By installing the cheaper touch-tone systems at the private home level and coupling the dial pulses to the electronic switching with a special converter, the cost could be reduced dramatically. INDETEL might have forgone a short-term profit from sale of the more expensive subsets, but by good system planning and adaptation, it protected its reputation and the confidence of the government.

Neither the customer nor the supplier could have evaluated independently the best course of technology development without rigorous planning and system management procedures. The use of the touch-tone device, converters, and electronic switching instead of the alternate approach will result in an overall saving of $75 per installation. As the infrastructure develops and manufacturing techniques are improved, the proposed system can be adapted to a complete tone-dialing integration due to the flexibility built into this initial planning.

Documentation and Manuals
Though the direct human exchanges and transfer of skills through training are the most critical type of technology transfer, the basis for all on-going manufacturing and adaptation is the existence of

a wide range of documentation. Design drawings for the telephone subset alone number over 10,000 and there are many commercial, functional, manufacturing, and technical specification manuals. Quality-control procedures, purchasing, accounting, personnel management, training techniques, industrial engineering and other techniques are all spelled out in special guides. These planning, management, and skill development manuals are important for on-going human resource improvement, but the crucial documentation for production are the specifications and engineering drawings.

Basic to any product development are the initial commercial specifications that come from the customer or government. In the past, the Mexican government has worked with the different private telecommunications firms in developing specifications to ensure that their guides are realistic and yet encourage the most advanced and effective technology.

With commercial specifications in hand, an organization like INDETEL works with its headquarters to seek out the best functional and technical specifications to meet the overall product objectives of the customer. Characteristics of operations, special features, and information on how the sub-assemblies will perform under given circumstances are the types of information contained in these specifications. The next level of written materials are the manufacturing specifications, which detail what raw materials are required, how to assemble, how to test, and how to finish surfaces. This information is required for each of the components and must be coupled with process descriptions detailing how to manufacture a given sub-assembly or element within the overall assembly. The process sheets, quality control, guide material lists, and manuals all form an essential library of working documents for production. Production of the Sonofon subset, for example, requires design input from the following sources:

Network design and production tooling for subset—Bell Telephone Manufacturing Co. (BTM) in Belgium; ITT Telecommunications plant in Corinth, Mississippi; Standard Electrik Lorenz AG (SEL) in Stuttgart, Germany;
Transmission and receiver capsules—BTM;
Tooling and design for dial mechanism—Fabrica Apparacchiature per Communicazioni Electtriche Standard S. p. A (FACE) in Milan Italy;
Mechanical elements—ITT plant in Corinth, Mississippi;
Semiconductors—Standard Electric Aktieselskab at Glostrup, Denmark.

In addition, ITT headquarters engineers also search for non-ITT components that might be better suited to the Mexican needs. For example, the special capsule magnets and related technologies are licensed to INDETEL from Phillips and Hitachi.

For the other product lines, INDETEL taps additional groups and houses within the ITT system. BTM, Compagnie Generale de Constructions Telephoniques (CGCT) at Paris, and Standard Electrica S.A. (SESA) at Madrid supply the documentation for general switching. PBX technology originates at BTM, SESA, and Compania Internacional de Telecomunicaciones y Electronica S.A. (CITESA) at Madrid and Malaga. The "glue that holds the whole system together," as one engineer put it, is the product of the integrating activities of the headquarters technical staffs and the on-going task forces, exchanges, and product conferences that ITT holds annually. The unique human resource of experienced engineers and managers is the critical element for creating the synergy of technology exchange between the widespread and divergent subsidiaries and associates of the ITT global system.

Human Resources as Technology

As we have seen, most of the underlying design technology is provided through documentation, but the critical stage of introducing the know-how and problem solving is left to personal contact and exchange. In production technology, the basic education of an engineer, technician, and manager must be complemented with both general industry experience and specific product knowledge. To master a technology, such as is requested in the dialing mechanism of a phone, requires initially a university degree in engineering. Added to that, four to five years of knowledge accumulated in the telecommunications industry, with five to ten years in the specific area of the dial mechanisms of a subscriber subset, are required. The ultimate refinement might be a specialization in the design and fabrication of gears within dial mechanisms for one particular type of subset. Without this extensive specialization and experience, the design technology cannot be implemented effectively.

The success of any large telecommunications company depends on its ability to gather together large numbers of widely varying specialties, which can then be used again and again for various operations around the world. Without the combination of the basic education, general industry experience, and product experience, the engineer cannot effectively master the specialty required to design, modify, and develop a particular product. The pyramid of these three requirements is manifested in the specialized engineering skills within the ITT headquarters technical staff and distributed throughout its different field operations.

ITT has organized itself so that each operating area and product line has a substantial amount of authority and independence for operating activities. To feed the global interchange and to provide technical input at the critical moment, the company has two large technical support groups located in Belgium and New York. These in turn are coordinated with regional product control centers and technical centers around the globe. These units consist of experts who make contributions toward solutions of various problems and coordinate technology for the product groups. ITT considers the integration and support these staffs provide as critical for developing new products, keeping exchanges and problem-solving dynamics, and delivering support when a unit encounters production, research, or distribution obstacles. The average experience of these staff engineers is over 23 years in their specialty. Among the many specialties represented in the New York and European technical staff groups are:

Engineering Management and Administration:
  R&D Administration, Engineering Management;
  R&D Costs, Budgeting, and Services.
Technical Directors Department:
  Food and Natural Resources:
  Automotive Products;
  Industrial Components;
  Applied Technology;
  Consumer Electronics;
  Environmental Technology;
  Telecommunications (switching, engineering standardization, digital systems, data and document equipment, computers, radio systems, communications cables, wire transmission, general transmission);
  Switching Systems (electronic switching, private switching, subscriber subsets, telex, electromechanical switching, electronic switching, switching maintenance);
  Defense Systems (avionics, military electronics, military communications).

The availability of over 300 specialists in the New York center and 200 in Europe's center lend great flexibility and dynamism to the ITT global technology system. In addition, ITT can draw on the engineers or specialists at any of its 186 manufacturing units with R&D. Their specialties are so widespread that a 133-page book is needed to list the different activity classifications.

The type of subset produced in Mexico is one of 32 subcategories under the private communication product line within the telecommunications group, which also includes switching, audio communications, data peripheral equipment, transmission systems, broadcast and mobile radio communications, transmission and central processing for data communications, and telecommunication supplies. The combination of all of the different skills within each subcategory of the product lines demonstrates the accumulated human resources that can be brought to bear on any problem or opportunity. During the evolution of the Mexican technology capability many of these unique skills were brought into play.

For example, one of the most difficult elements for the Mexican plant to master was the production of carbon transmitter capsules. Many considered this element in the realm of "black art" because it is impossible to qualify the exact conditions required for successful production, and some units in the ITT system had not succeeded in manufacturing this subcomponent. By the late 1960s, the Mexican engineers and technicians were anxious to extend their integration into this complicated area but had been discouraged from doing so by ITT headquarters. By coincidence, however, headquarters had dispatched a specialist in carbon transmitters to Mexico to work on a related problem, and his work with the local technicians over a six-month period enabled INDETEL to identify and correct the problems that were standing in the way of successful capsule production.

Initially the delicate carbon pieces were being ruined by contamination of workers who had eaten citrus fruits for lunch, and female assemblers who had certain skin characteristics, as well as by inappropriate cleaning and assembly procedures. All of these bizarre problems could not be tested for, or identified in manuals, but they could be analyzed by a specialist who had spent most of his life concentrating on aspects of carbon transmitter capsules. Because INDETEL had access to his skill, it now produces the capsules for its own use and for export, and they are of the very highest quality. The production in Mexico follows no set rules, relying instead on a method uniquely tailored to local conditions and equipment, developed from the combined experience of the specialist and the local capability of the Mexican engineers.

Transferring the Human Resources Technology
The specialized experience of the human resources of ITT represents the critical technology capability. The use of exchanges, global conferences and task forces are the means by which the corporation keeps the technology flowing and creates the continuous

contact and interface. At least two major task forces a year are held on station apparatus such as subsets. In 1973 one task force met at Corinth, Mississippi, to visit the plant and examine new tools and machines, as well as to participate in the normal discussions, presentations, and group problem-solving exercises.

Task forces generally cover two days with side meetings that last well into the evening. At the meetings progress reports are given by each company on its main projects so that attendees are current on products being developed throughout the ITT system. Problems, some commercial but mainly technical, are discussed. The result is either a decision on the design direction to be followed, or an action program to obtain more data on which to base a decision. Usually the solution is entirely satisfactory to the unit with the problem but sometimes decision is reached that is best for the overall corporation, but which may be considered second-best by the individual unit.

For example, the new transducer for electronic telephones can be tooled by every major plant. Other techniques sought at task forces by competing units were not fully adaptable around the world. But headquarters required a standardized approach to the development of this important transmission/receiver capsule, which has reduced size and interchangeable characteristics.

In addition to exchanging information, task force participants have the opportunity to meet their counterparts from other ITT companies. This personal contact facilitates subsequent information flow between ITT companies throughout the world.

Managers are encouraged to visit different plants before buying equipment or installing new techniques. If they opt for equipment already in use elsewhere, their engineers and technicians are sent to that plant to train alongside experienced operators. A plant installing new equipment can ask for expert help in installation, use, and specific problem solving. In 1974, two skilled tool makers from CITESA were used at INDETEL to train the Mexicans in specialized maintenance and operations of tool and dye fabrication machines. INDETEL paid only the salaries and expenses of these specialists during their stay in Mexico.

On the production line, the Mexicans significantly modified the welding tables developed at BTM. In Europe, tests had indicated that the tables should be covered with grey material for eye strain relief, fans placed in the table to remove fumes, and certain types of chairs utilized to reduce strain. In Mexico, the assemblers felt more comfortable with green material, complained that the fan blew in their faces too much, and modified the chair due to its discomfort. Even the simplest industrial solutions are often radically transformed to be effective in a transfer from one culture to another. Psychological concepts of color, types of tools used, language and even attitudes toward supervision and group relationships must be considered in transferring know-how.

The effectiveness of exchanges and on-the-job-training is sometimes reduced by unexpected cultural and attitudinal obstacles, of course. Diminished effectiveness of seminars and task forces because of language problems is a common example. Managers in Mexico claim that the availability of ITT manuals in Spanish has helped enormously, but that many of the technical visits are not carried out in Spanish which minimizes the value of the information transfer.

Cultural biases can also impede transferral of technical expertise. At one point, for example, INDETEL purchased a specialized process from a firm in Switzerland. A team of Swiss experts went to Mexico to install the equipment and train Mexican workers in its use. Initially, the training went smoothly, but later, when the Mexicans attempted to run the machine under the Swiss direction, the output was unsatisfactory. The Swiss technicians instructed the Mexicans just how to operate the device and told them that if anything went wrong not to attempt repair or adjustment, but to seek engineering aid. Several attempts by the Mexican workers at operating the machine under these conditions failed.

At this point in the program a long weekend provided a break in the tension. During the holidays the Mexican foreman and his workers went to the plant alone, ran the machine, carried out modifications, and corrected several small breakdowns without assistance. When the plant reopened after the holidays, the Swiss were amazed to find that the Mexicans, when left alone with complete authority, could run, adjust, and modify the device successfully. The attitude of the foreign specialists during training had apparently frustrated the workers, and the weekend had given the Mexicans an opportunity to prove that their engineering and technical skills had been underestimated by the Swiss.

Many examples exist of how overly standardized techniques can inhibit utilization of technology, so INDETEL engineers and professionals continually encourage the skilled workers not to merely carry out assigned tasks, but to step outside their job descriptions and experiment with new procedures. It was through this type of encouragement, for instance, that INDETEL engineers learned that the technicians they frequently had been calling in to repair a finicky compressor were unnecessary: the job could be done in-house by several of the workers who simply needed to be given the chance to assert themselves.

The transfer of knowledge, then is only part of the job. The encouragement of utilization and development is also a difficult part of the transfer process. Materials on organizational behavior and group behavior are now a standard part of ITT's global training to help managers relate to these types of psychological and sociological problems. Unfortunately, one of the drawbacks of the highly skilled expert staff is that it cannot possibly contain all of the language and cultural skills necessary to ensure full exploitation on a global basis of the experience and know-how it embodies.

Training of Direct and Indirect Labor

To date, INDETEL has trained more than 300 electrical and industrial engineers in the United States, Europe, and South America. It is difficult to estimate exact time required for such training since the education and development continues indefinitely through joint work programs, exchanges with other plants and centers, and exposure to new processes, designs, and procedures. Quantification of this type of absorption of technology is difficult and for any individual it might mean a gain of 10 to 200 formal days of training and an on-the-job training impact of 20 percent of their work for three to five years (250 × 3 × .2 = 150 equivalent man-days). Examples of engineer training are as follows:

Quality Control Engineer:
   60 man days of full-time formal training;
   72 man days of on-the-job-training presuming that actual learning
      impact = 10 percent of each OJT day.
Industrial Engineer:
   120 days of full-time formal training;
   144 man days of on-the-job-training presuming that actual learning
      impact = 30 percent of each OJT day.
Production Control Engineer:
   240 days of full-time formal training;
   162 man days of on-the-job training presuming that actual learning
      impact = 30 percent of each OJT day.

After several years of operation, a company like INDETEL builds years of unique experience that is then transferrable to related industries. Much of the high turnover at INDETEL results from vigorous competition for the trained engineers and experts. Presently from 20 percent to 15 percent of engineers, technicians, and production clerks leave the company each year. This turnover requires continuous training on the order of 30,000 man-days a year to maintain indirect labor capabilities. The same phenomenon can be identified in direct labor, with turnovers of between 5 percent and 18 percent requiring 7,400 man-days of training a year to maintain skill levels. The actual number of equivalent man days of formal and OJT training varies with individuals and stages of development in the plant.

Tables 6-3 and 6-4 illustrate the dimensions of training needed to cover continuous turnover and introduction of new products. Including the days spent traveling each year to train or exchange information with other units, an engineer during his first year in quality control would receive 118 days of equivalent training (formal OJT, trips 60- 72/1.5 + 10), 34 in his second year, and about 10 full days a year from then on through corporate exchanges, task forces, and seminars. This does not include the outside training he is encouraged to pursue by the corporation or the continuous accumulation of knowledge of new techniques and procedures. The net training contribution of the company is estimated on what it would require to prepare an engineer, clerk, and so forth to do his initial task at a satisfactory level. To calculate the value of ongoing training would be even more subjective than the OJT measurements, but it is substantial and critical for the development of the specialized industry and product capabilities.

To gain an appreciation of the training and exposure required, to establish a production technology with the capabilities of INDETEL, we can sum the training required for the different types of total direct and total indirect skills: 15,960 man-days for supervisors, 6,780 for foremen, 41,692 for technicians, 41,260 for engineers, 10,000 for production clerks, and 2,552 for management and executives. This totals 118,244 man-days or 472 man-years of training in indirect labor. For direct labor, 79,540 man-days or 317 man-years of training would be needed. If all the training materials, experienced trainees, and specialists could be assembled to recreate the accumulated training and capabilities of INDETEL in 1974, they would represent a training transfer of some 657 man-years of training for 1,049 indirect employees and 1,710 direct laborers. This type of estimate assumes that all the correct specialties are readily available and a system exists that can support the training, provide the equipment and installations to train on, and manifest an industrial identity suitable to attract and motivate the individuals.

Between 1968 and 1973 INDETEL spent over $4 million on formal vestibule (classroom) training. This figure does not represent

Table 6-3.   INDETEL Indirect Labor, Training, Turnover, OJT, and Initial Experience Required, 1974

	Numbers	Turnover %	Turnover #	Experience Required Ed.	Experience Required Yrs. Exp.	Estimated Days of ITT Mexico Training (Days) Formal	OJT	Average Days on Foreign Trips or Visits Per Cap	Total	Estimated Turnover Training Man Days
SUPERVISORS										
Prod.	70	.20	14	HS	1.5	60	54			1,596
	70	.20	14	HS	1.5	60	54			1,596
Install.	57	.05	3	HS+	1.5	240	144			1,152
FOREMEN										
Prod.	15	.05	1	HS+	1.5	100	96			196
Install.	8	.05	0	HS+	1.5	240	240			
TECHNICIANS										
QC	71	.25	18	HS+	.5	60	72			2,376
IE	14	.25	4	HS+	.5	80	144			896
Eng. Dept.	175	.25	44	HS+	.5	120	96			9,504
ENGINEERS										
QC	27	.25	7	BA	1	60	72	10	270	924
IE	38	.25	10	BA	1.5	80	144	20	760	2,240
PC	114	.25	29	BA	1	160	96	20	1,140	7,424
CLERKS										
QC	)			HS-	.5	5	12			
IE	)			HS-	.5	20	48			
PM	)314	.20	63	HS+	2	2	6			2,000
Fin	)			HS+	2	20	12			
Sec	)			HS	.5	5	6			
MANAGEMENT	132	.10	13	BA+	3	30	12	15	990	546
EXECUTIVE	14	.25	3	BA+	3	60	12	80	560	216
TOTAL	1,049		209						3,720	30,666

TRAINING ESTIMATED TO REPLACE TOTAL PLANT INDIRECT LABOR   118,244 man days
or
472 man years

Table 6-4. INDETEL Direct Labor, Training, Turnover, OJT, and Initial Experience Required, 1974

	Numbers	Turnover		Experience Required		Estimated Days of ITT-Mexico Training (Days)		Estimated Turnover Training Man Days
		%	#	Ed.	Yrs. Exp.	Formal	OJT	
PRODUCTION OPERATORS								
Normal	1,023	.18	184	HS-	0	25	6	5,704
Special	40	.05	2	HS-	2	10	12	44
INSTALLERS	217	.05	11	HS-	0	20	96	1,276
TESTERS	175	.01	2	HS+	1	20	96	232
AUXILIARY	255	.12	31	HS-	0	5	0	155
HELPERS				0	0	0	0	
TOTAL	1,710							7,411

TRAINING ESTIMATED TO REPLACE
TOTAL PLANT DIRECT LABOR   79,540 man days
                                                     or   317 man years

the time or cost involved in the OJT or direct exchanges, nor the cost involved in providing the training materials, manuals, and so forth from ITT headquarters or other units. Table 6–5 shows that for the year 1974 vestibule training will represent $914,000 with $309,000 involved just to maintain the training department. An important focus of the classroom training is to initiate the overall training effort of direct and indirect labor and of customers.

Training for Customers and Students

In addition to producing and maintaining the several million subsets utilized in Mexico, INDETEL is responsible for aiding in development of telecommunications technology for the government and training government technicians and engineers. Since the major responsibility for regulating the vast telecommunication network belongs to the government, its personnel must be fully acquainted with the design, maintenance, and operating characteristics of the subsets, switching, transmission, and PBX systems. In 1974 INDETEL spent $58,000 for basic vestibule training of government personnel and delivered approximately 11,400 man-days of formal and on-the-job training.

Table 6-5. INDETEL cost of Vestibule Training, 1974 ($000)

AREA	*Direct Labor*	*Indirect Labor*	*Overhead for Training Department*	*Customer Training*
Prod. workers & technicians	187	135	147	
Maintenance & installation	201	8	66	58
General office	—	16	38	
Engineering	—	—	58	
Total	388	159	309	58
GRAND TOTAL	914			

In addition to the customer training with the government, INDETEL provides formal programs for training student technicians and engineers on its equipment. During 1974, the 40 individuals involved in this program received over 800 man-days of training. Together, the government and student training programs are a major technology contribution totalling 11,400 man-days of training as seen in Table 6-6.

The type of training represented by the vestibule lectures and the OJT of INDETEL moves the engineers, technicians, and clerks from the realm of basic education to that of specialized industrial know-how. Student trainees are encouraged and guided by the government to supplement their initial foundations with actual experience on advanced products and industrial techniques in their disciplines. Since Mexico's rate of absorption of technology is constrained by the capability of the infrastructure to provide technicians, tool makers, engineers with specific product experience, and other specialized skilled workers, the on-going training programs of an operation like INDETEL represent a continuous technology transfer to the nation as a whole. Considering only the training lost in annual turnover and the outside work with customers and students, INDETEL contributes 49,477 man days per year of specialized telecommunications know-how to the external environment. Additional

Table 6-6. INDETEL Customer and Student Training, 1974

	Numbers	Days of ITTM Training Formal & OJT	Total Training Days
GOVERNMENT TECHNICIANS			
Subsets	200	5	1,000
Switching	20	10	200
GOVERNMENT ENGINEERS			
Switching	100	80	8,000
Transmissions	15	40	600
Private Switching	20	40	800
STUDENTS	20	40	800
	375		11,400

infusions of know-how into Mexico are provided by INDETEL's on-going work with suppliers and local associations.

Direct Exchanges

During the crucial period of new development or production change, large groups may be exchanged in several directions, as was the case when the group of engineers were placed in CITESA, each for six months to a year to acquire the basis for Pentaconta switching. Similarly, two skilled workers and a foreman were placed in the same plant to receive direct experience in maintenance of specialized machinery. These unique transfers are an integral part of the direct contact training and exposure. On a more regular basis the senior staff spends 75 to 100 days a year visiting other plants and working out common management problems. The professionals and engineers travel less frequently but average 10 to 50 days per man outside of the plant with other experts on special problems, seminars, or workshops.

Most of the exchange focus during the 1950s was on integration of subset production. By 1961, technology focus was moving to switching. To establish the initial switching technology, six engineers and production experts were each sent to Belgium during 1963 and 1964 for six months to a year. This direct exposure formed the basis for the buildup in the general and private switching operations. From eight engineers in 1965, INDETEL expanded its staff to include 180 engineers and technicians in this one product line by 1974. The technical base and financial cost of this technology expansion was supported through the profit and know-how resulting from the subset activities. Local profitability and technology growth in one product line supported the expansion of others.

INDETEL receives four to five major visits from the New York headquarters staff on subset technology and in addition hosts three to four visits from other Latin American operating centers. At present, there are several joint development programs being carried out in cooperation between the Latin American subsidiaries, supported by the global R&D planning and programming of ITT. From Europe, INDETEL can expect at least two visits annually from each of the large telecommunications centers at Spain, Belgium, and Germany.

The course such visits take is fairly standard. For example, if New York technical staff in the station apparatus field were to visit INDETEL, progress on all new projects would be evaluated first. This would enable any deficiencies in INDETEL to be identified. Next the staff would concentrate on locating the best solution to the problem. In some cases the initial request would originate from Mexico; in others, the headquarters might spot the need for help and move on its own accord to supply information and recommendations.

During one visit it was found that the hook switch used by INDETEL was becoming unreliable. This was traced to worn-out tools. Rather than reproduce tools for this hook switch, the visiting technicians suggested a new and better one developed by CITESA, which was ultimately adopted.

A problem of flow-soldering this new hook switch onto the circuit board was identified on a subsequent visit. It was costing INDETEL $0.15 per subset to hand-solder the unit after the rest of the parts were flow-soldered into place. By coordination with CITESA, which did not have this problem, it was found that the INDETEL flow-soldering equipment was not operating properly due to temperature fluctuations and changes in flux. When this was corrected, the hook switch was then found to flow-solder properly into place.

Another example involves the secretarial set made by INDETEL, which was identified as a high field maintenance item because of the complicated nature of the various switches. INDETEL made a visit to Corinth, where a number of suggestions for improvement were made. These were implemented by INDETEL with highly satisfactory results.

In all these cases, a problem was identified by INDETEL, headquarters suggested methods to arrive at the best solution, and an agreed course of action was decided upon by INDETEL. Note that in most of these cases, neither CITESA nor Corinth had anything to gain by supply of assistance. Without headquarter coordination it would probably not have taken place.

The Mexican managers indicate that it is easy in this type of specialized production to succumb to inertia and push for minor refinement of existing products rather than consider new products involving a certain amount of risk. But for INDETEL the natural tendency to stagnate is countered by the influence of the global ITT system and the continuous exchanges, which work to maintain an open and dynamic approach to engineering and research. The network, stimulated by the cross-fertilization, encourages direct access to any part of the system since the personal and professional identities have been established in earlier working groups and task forces.

Tests, Equipment and Specialized Technologies

INDETEL incurs almost 70 percent of its production costs in purchases of raw materials and components. Supplies range from screws, plastic powders, electric and wire cord to bulk steel, brass, and rubber piece parts. The early development of effective stores management and purchasing control was critical to profitability and production management. ITT experts helped in the planning and provided technical support for the development of a global purchasing capability at INDETEL. Great difficulty was encountered in developing importing procedures for critical components and this received on-going attention from the complex purchasing and supplies planning activity. At present, purchasing is handled by a full-time Mexican manager and a small staff in Belgium with BTM that maintains contact with global vendors and monitors developments in European markets.

One of INDETEL's major activities is the repair and maintenance of installed equipment. Revenues from repairs is almost equal to that of production, although the profitability in this area is much lower. There is considerable emphasis on developing effective materials and durable assemblies to reduce needed maintenance. Special diagnostic and maintenance procedures are developed between Corinth, FACE, BTM, and INDETEL to extend the life of the subset.

During production the different components are tested at approximately sixteen different sites using a variety of tests and devices. Specially adapted oscilloscopes, acoustic metering devices, and a wide variety of current and signal evaluators are employed. These tests and equipment are introduced as different levels of production sophistication are reached. At certain points the Mexican system has to tailor the basic equipment and procedures to its own specific requirements.

While introducing the production of transmitter capsules, the Mexican engineers developed a unique acoustical testing device with the help of a visiting ITT scientist. When the test was first introduced, it caused significant problems for production control, but ultimately the joint involvement of the acoustical specialist and the industrial engineers modified the test so that it could be used to improve the accuracy of measuring sensitivity without interrupting production flows.

The accumulated testing equipment at INDETEL represents one of the most advanced concentrations of hardware of its type in the country. The Mexican government uses the facilities extensively and relies heavily on the INDETEL staff for new innovations in tests and standards. Present equipment inventory for testing is valued in excess of $1 million. When Tel-Mex recently opened its own laboratory, ITT technicians supplied layout information and standards for equipment to be purchased. Vendors and local contractors are also encouraged to use the test equipment and laboratory facilities to improve their own quality control and standards.

As the Mexican telecommunication industry evolves, it is the cooperation between the government, local, and various foreign technology suppliers that ensures better standards and improved testing and evaluation procedures.

PRODUCT IMPROVEMENT AND THE DEVELOPMENT LABORATORY

During the 1960s, with the addition of new products and expansion of markets, INDETEL began placing greater emphasis on securing and implementing new technologies. By 1972 the company had developed a centralized development group and laboratory with the specific mission of providing technical and R&D support to all the product lines. The newly formed group acted as an interface between the engineering of INDETEL and the corresponding levels of the foreign units. In the words of the ITT Mexico plan, the development laboratory group would become an "on-job-training and re-training center for both new technologies" and existing products. This highly coordinated technology center would assure the continuing resources to improve self-sufficiency, growth, and appropriate product development.

Product Modification and Development

For the new generation of the Sonofon, the Mexican government has requested significant style changes. Two years of circuitry design work will be needed from SEL on internal design, while Mexico will provide one full-time engineer with six-to-eight years experience, and a support technician to respond to the requirements for reworking the housing and the related accoustical and electronic modifications.

Sonofon I was developed about 1969–70 and intended as a new style of set to replace a shape similar to the 500 set still in use in the United States and Canada. The shape was the result of early industrial design work done for Corinth but not introduced by them in the United States. The tools for this design were readily available for transfer from Corinth to Mexico. The initial transmission circuit of the set was designed by BTM. With a little help from BTM and STL, it was modified to fit the acoustics of the new Sonofon shape.

Later, as Sonofon I tools began to wear, it became apparent that a new set should include some of the advances made in the ITT system. Sonofon II was designed with the assistance of SEL. This new design introduced in 1974 allows reduction of cost, easier replacement of parts, more versatility to adapt the set to various low production volume requirements without major additional tooling costs, and so forth.

The ringer now made by INDETEL is also being redesigned. The present high cost of this component is partially due to the price paid for the coil from an outside supplier in Mexico and partially because the design licensed from another firm is inherently high cost. New ringers have been designed by ITT and one of the best was chosen with help from headquarters staff. The INDETEL designers have used the new input to develop a rough lab model of the modified design to replace the present device. Further work at INDETEL will refine this component and allow both a cost reduction and a product improvement.

The present development of the "hands free" subset is another example of how the best European technology from FACE is selected by INDETEL to be modified locally to meet Mexican standards. This pattern has been repeated in the importation of all of the switching and transmission technology, with the design contribution of the Mexican engineers growing in importance. In the near future, Mexican specialists will be using technology from ITT operations in Norway and Sweden to initiate products for rural telephone switching controls to improve the service of the central exchanges. The new laboratory will centralize the flow of these foreign inputs and tailor them to meet Mexican requirements.

Development Laboratory

During 1972, the development laboratory was formed to begin centralizing the wide variety of technology transfer activities. The increasing technology base begun with subset assembly, and expanding through production of subsets, switching and transmission had reached a point where efforts and resources could be economically directed at a new centralized R&D effort. Technical service, technical assistance, training and equipment development were major missions for the new lab, together with maintaining coordination with INDETEL's quality control and manufacturing departments. The laboratory also provides an effective interface for the company with ITT headquarters technicians and the European technology sources.

To accomplish these goals the laboratory immediately began increasing its manpower experience level, with emphasis on specialized engineering. As shown in Table 6-7, by 1978 the lab hopes to increase its size from the original 12 to 72 engineers and technicians with an average experience of 6.6 years as opposed to the 5.3 year average at the beginning. The initial training and staffing during 1973 to 1975 would keep the average experience level lower overall, but initiate a base of trainees for experience development, since these specialized skills could not be effectively recruited from the outside.

Table 6-7. INDETEL Development Lab Present and Planned Staff, 1972–1978

	1972	*1973*	*1974*	*1975*	*1976*	*1977*	*1978*
Engineers and Technicians	12	20–28	44	50	56	64	72
Trainees	20	12–10	6	6	8	19	11
% Trainees of Lab Group	166%	60–35%	14%	12%	14%	14%	15%
Average Eng. Experience	5.3	4.6–5.1	4.7	5.6	6.1	6.3	6.6

The build-up in experience would be coupled with a budget build-up for computer systems, measurement equipment, test equipment, and other laboratory and office hardware, totalling $650,000 between 1974 and 1978. INDETEL's overall budget for this centralized technology transfer unit, was $500,000 in 1974 and may be $1 million by the end of 1977.

The allocation of laboratory use reflects the development needs of INDETEL: about 80 percent of the laboratory projects are directed at switching activities; 10 percent are keyed to supporting the subset improvements; another 10 percent goes to transmission and PBX activities. The imbalance is due to the fact that the product lines of subsets and transmission have their own engineering support, whereas the new switching activities receive almost all their technical support directly from the lab, which acts as a training ground for the future product line technicians and engineers.

At present the laboratory staff is actively creating its engineering skill base in Pentaconta switching and is conducting feasibility studies for introduction of the system. The staff also provides training for activities such as computer-based maintenance, tool-ticketing equipment, quality and concentration tests, and a wide variety of value engineering activities on present production activities. Besides the training, feasibility studies, and value engineering activities, the laboratory carries out an extensive array of research and development on multi-frequency circuits, rural electronic switching, electronic test equipment, computer-aided design techniques computer-based installation testing, and specialized robot supervision devised to control certain production. Much of this R&D is directly financed by headquarters.

The development lab is also responsible for establishing standards for the locally manufactured or available components used in INDETEL production. In addition, much of the on-going work on quality control, engineering, and production aid to suppliers is directed out of the central laboratory and development group.

The laboratory and development group also works closely with Mexico's Ministry of Industry and Commerce on national standards for telecommunication equipment and participates with working groups of the National Committee of Telephones (CCNNTEL). Laboratory staff spends extensive time with the government on joint work groups to evaluate components, develop quality tests, and provide technical counsel. The cooperative programs established by the government ensure the most effective homogeneous standards for the national telecommunications system.

INDETEL is now a major center for development, research, and engineering for the global system as the outputs and results of these new efforts will not only serve the local plant, but will provide new techniques, procedures, and products for the whole system. The success of the new laboratory and product development group demonstrates how the technology evolution, properly controlled and matched to a growing infrastructure, can change from simple assembly to advanced research and development. INDETEL reached its local content objectives by the 1960s and its technological independence by the early 1970s. By the latter half of the '70s, it plans to represent a major technology center for ITT activities in Latin America (training and development) and a specialized research and testing site for ITT telecommunications products.

Dissemination of Technology

The formation of the laboratory and development group accomplished two objectives: (1) centralization of technology transfer to the INDETEL organization; and (2) coordination of the on-going dissemination of the technology from INDETEL to the external environment. In addition to the training of students and government specialists, the INDETEL staff works closely with suppliers to develop necessary skills, quality, and new components for production needs. In some cases the Mexican suppliers simply copy the designs supplied by ITT, but in others close supervision and training is required.

In one case, INDETEL had great difficulty developing the proper elasticity in its handset cords. Too much spring in the handset could pull the subset off tables or desks; too little made the cord a nuisance. Since 30 percent of the repairs are related to cord or grommet breakage and separation, the ability of the supplier to deliver the best quality of cord was critical.

To overcome the interrelated problems of plasticizing different coatings, correct wire tensil, and durability of grommet and cord, ITT headquarters supplied information on different formulas for the plastic and wires, and specialists in mixing rubber compounds and forming the grommets were dispatched to see the supplier. The supplier was then sent to BTM to observe cord manufacture and testing and was able to overcome the problem by development of a satisfactorily "springy" cord. This illustrates a problem involved in improving many subcomponents -that is, overcoming the resistance of protected Mexican suppliers who don't have to face normal competition.

Considering that INDETEL has over 200 suppliers and many have to be individually developed, like the cord vendor or the plastics fabricator who had to visit Italy for training, the impact of INDETEL specifications, training, and procedures is quite extensive. From the involvement with ITT–Mexico the supplier may develop his own specialized product engineers, mold makers, and tool and dye craftsmen.

Specialists have to be trained for quality control, chemicals handling, and product testing to meet the rigid standards. The result of this involvement often means new capabilities to support other emerging Mexican industries such as the production of household appliances, automotive electronics, and commercial goods. The demand and technical support created by INDETEL and other telecommunications companies provide a critical element of local scientific and industrial infrastructure evolution.

INDETEL UNDER LOCAL OWNERSHIP AND NATIONAL TECHNOLOGY GUIDELINES

In December of 1972, the government of Mexico enacted a law regulating the transfer of technology into the country. The stated purpose of this legislation was to implement the UNCTAD objectives for establishing institutions for the specific purpose of dealing with issues of technology. The law required registration and review of all licenses covering patents, management contracts, and trademark agreements. A special technology agency, the National Registry of Technology transfer, was established to evaluate technology flows. The chief of registry acknowledged that they had not devised an effective and comprehensive methodology for determining relative benefits and costs of an individual transaction or investment.

However, the Registry does follow certain guidelines for its negotiations and decisions on technology and attempts to eliminate the following types of undesirable technology transfers:

When the technology is freely available in the country;
When the price paid is excessive;
When managerial control is given to licensor;
When there is obligation to return improvements or inventions free of charge to the licensor;
When there is an imposition or limitation on R&D undertaken by the licensee;
When there is an obligation to purchase equipment and raw materials from the licensor;
When there is a restriction in respect to exports contrary to the national interest;
When the use of complementary technology is prohibited.

The law encompasses several other situations and restrictions, but the above list identifies how the government will intervene to judge and rule on technology relations of INDETEL and the ITT world system.

Prior to 1974, INDETEL participated in the ITT global technology system as a subsidiary—we might call it a full and internal member. Much of the technology flowed freely as a natural part of business operations. As noted, the critical components were represented by exchanges of personnel, specialized training, and availability of basic documentation, reports, and specialized support staff. In 1975, ITT's ownership dropped to a minority of INDETEL shares. It would be instructive to determine later what impact this had on in the technology relationships.

ITT entered into an agreement with the government that provided for Mexicanization of the INDETEL telecommunications equipment through ownership of 51 percent by the Mexican government and nationals. The government actually owns 17 percent of the company, with 11 percent being transferred from International Standard Electric Corporation (ISEC), a wholly-owned ITT subsidiary, to Banco de Mexico as trustee. The Mexican state credit organization, SOMEX, has an option to purchase these shares, or they can be gradually fed into the private capital market. Individual Mexican investors own 23 percent of the company. ITT now owns 49 percent of the firm, where previously it held 60 percent ownership in INDETEL.

The reduction of ownership control in INDETEL is in line with the government's policy of increasing Mexican control and direction of critical industries. Many of the government technicians and planners believe that their involvement and monitoring of the previously foreign-dominated technology flows will lead to overall improvement within the scientific and technological infrastructure of Mexico, as well as reduce the costs of technology transfers. ITT's policy has been to cooperate with this direction and to continue supplying technology through separate agreements, which are now necessary in their minority role. Headquarters staff at ITT believe they can maintain the quality of technology and support that has developed over the past twenty years.

At the local level, the reaction is mixed. Most of the local managers and engineers are enthusiastic about the Mexicanization concept, but dubious about their ability to gain the ready support and aid as in the past. One engineer stated that "the government doesn't understand how important it is to be able to telex a problem directly to a centralized staff like the one in New York and get support back immediately on any type of production, design, or development problem." As the INDETEL organization is formally separated from the normal ITT system, it will naturally become more difficult to freely circulate or gain access to the different technology resources.

"I used to be able to go to any plant or headquarters and walk around asking questions, looking at files, or observing but now it may be more difficult as ITT does not have the security of control that it used to have," is the feeling of one INDETEL staff officer. The New York headquarters staff is confident the new arrangements will allow continuous flow and support, but the rapidly changing dynamics of telecommunications technology and the need for various ways of complete exchange and exposure force the question regarding the ultimate effectiveness of this new system of technology relationships. The expanding lab and new product focus can only function optimally if continuously cross-fertilized.

To avoid diminishing the technology flows to INDETEL now that it is not controlled by ITT, the company has entered into a series of agreements with the corporate headquarters at ITT. The contracts and convenants set up guidelines for technology access and payment for these services.

While the global system provides access and economies of scales for smaller organizations regarding technology resources, the new national programs draw attention to particular local needs. The intervention of the government in problems such as global allocation or research, equipment and raw material requirements, and exports of new developments and products challenges the effectiveness of the global technology pool and coordination of the large technology producers. The cost is unknown at the moment, but it is hoped that better local integration will be gained and technology flows overall will relate more directly to the perceived social and economic goals of the country.

If there is a natural pattern of technological development linked to market, infrastructure, and source of technology, then it may be possible to intervene in the process at a critical stage to speed up or improve the transfer. However, it does seem clear that in a situation like INDETEL, the creation of product development and research could only follow the careful build-up of technical and engineering skills in manufacturing. Continuous outside support was necessary to remain competitive and to introduce new technologies. It is difficult to evaluate the cost or opportunity loss involved when the free access and continuous exchange items are replaced by formal contract provisions. However, we can say that certain fiscal policies relating to the elements of market and infrastructure might be more accurately developed as opposed to the intervention and restructuring of technology flows, for which we have so little understanding of the actual methods and critical factors relating to securing and transferring the appropriate know-how.

Part III

Pfizer Corporation

Chapter 7

Pfizer–International

Pfizer, Inc., was founded as a partnership in 1849 by Charles Pfizer and Charles F. Erhart, both recent emigrants to the United States from the town of Ludwigsburg in the Kingdom of Wurttemberg, Germany. The partnership was named Charles Pfizer & Co. with Pfizer, then 25 years of age, as the senior partner.

Pfizer, a chemist, and Erhart, trained as a confectioner, envisioned their firm's mission as that of a manufacturer of chemicals not then commonly produced in the United States. With this objective, the partners purchased a small building in Brooklyn, New York, for the production of its first chemical, santonin. The successful manufacture of this product, widely used to compound vermifuge medicines, was followed by the production of iodine preparations as an added service to the wholesale and retail druggist.

By 1855, mercurials (including calomel), borax, boric acid, and refined camphor were added to the Pfizer production list. In 1862, the partnership initiated the manufacture in this country of tartaric acid and cream of tartar. In 1880, the firm began producing citric acid from citrate of lime imported from Italy. This method of production was continued until the early 1920s when Pfizer researchers achieved a major breakthrough in fermentation chemistry by perfecting a process to transform ordinary sugar into citric acid. This expertise in fermentation chemistry was to play a key role in Pfizer's leadership during World War II in the mass production of penicillin.

Shortly after that Pfizer breakthrough, the company became one of the early manufacturers of the antibiotics streptomycin, bacitracin, and polymyxin. In the meantime, Pfizer's fermentation technology was brought into play during the 1930s to establish the company as a pioneer in the production of such important chemicals as gluconic acid, itaconic acid, and fumaric acid. Also in this era, Pfizer was among the first companies to begin research on an industrial method of synthesizing the vitamin asorbic acid. A successful conclusion to this research resulted in the commercial production and sale of vitamin C by Pfizer in 1936. The Pfizer production of riboflavin and other vitamin products followed.

In Pfizer's more recent history, the company's manufacturing expertise has been combined with innovative research and marketing programs and has resulted in the transformation of the company from a small domestic concern with sales of $61 million in 1950 to a diversified multinational corporation. This period of rapid growth was begun in the early 1950s with the company's entry into the pharmaceutical marketing field with the introduction of Pfizer-discovered Terramycin, its rapid expansion into markets abroad, and its diversification into animal health and other agricultural products. By the late 1950s, Pfizer pharmaceuticals, bulk chemicals, and animal health formulations were sold throughout the world.

In the early 1960s, Pfizer entered the consumer products market on a full scale with Leeming, Pacquin, and Coty products. The company also entered the field of materials science in the 1960s with iron oxide, lime, limestone, and specialty metals. More recently, entry has been made into the diagnostic, prosthetic, medical specialties, and dental health care areas as well as animal and seed genetics.

Today, the Pfizer name stands for a diversified line of quality products sold the world over. Pfizer's far-reaching operations are conducted by some 40,000 men and women employed in a global organization with facilities that include a network of 128 production units in 39 countries.

Grouped into five major product areas, Pfizer ethical pharmaceuticals and health care products (48 percent of total sales), chemicals (16 percent), agricultural products (14 percent), consumer products (14 percent), and materials science products (8 percent) serve the world's health team, a variety of industries, consumers, and farm communities in over 100 countries. Annual sales are in excess of $1.5 billion.

The majority of Pfizer's pharmaceutical and health care activities are in the field of chemotherapeutic agents for a wide variety of human illnesses. Pfizer's historical leadership in penicillin and other antibiotics established a base for developments in psychotherapeutics, antidiabetics, cardiovascular drugs, and other medical specialties. Dental items are also sold on a worldwide basis with diagnostic and microbiological products produced primarily for the U.S. market. The remainder of Pfizer's production is divided among the chemical, consumer, agricultural, and minerals areas.

Pfizer chemical products are utilized by food, beverage, industrial, and other pharmaceutical manufacturers. Bulk production of (citric acid), fumaric acid, ascorbic acid, sodium gluconate acid and so forth is distributed throughout the world. Within the consumer product area, Pfizer markets under the Coty, Leeming, and Pacquin brands names. Consumer products include toiletries, powders, ointments, and other nonprescription health items. Pfizer's Agricultural Products Division produces a broad range of animal health and veterinary formulation, as well as vitamin and feed supplements. The materials science products include specialized iron and chrome oxides, refractory specialties, lime limestone, and other mineral, pigment and metal products.

GENERAL ORGANIZATION

To deal with such a wide product and area diversification, Pfizer International is organized on a geographic basis. For daily operations and support, each facility can rely upon a centralized headquarter's staff and several technical centers to provide specialized technology and management assistance. The various staff and field operations report to one executive vice president, one operations vice president, and the president. Regional support is supplied by a technological center. Additional support comes from headquarter specialists, as well as other laboratory or staff units throughout the world. Pfizer relies strongly on its country and regional managerial system for coordination. However, there are specialized reporting procedures for quality control, pharmaceutical, medical, and agricultural development to insure quality and uniformity throughout the company. The effective control of production and product development activities is a paramount objective of the company, superceded only by its commitment to absolute quality control throughout the world. Multiple reporting lines and rigorous standards are considered critical to Pfizer trademark protection, as physician and consumer confidence is a principal advantage and responsibility for the company.

Similar to many firms with a wide variety of technology exchanges and multiple products, Pfizer decentralizes certain management and product functions, but centralizes its technology and specialized skills in regional centers for effective global application. The specialized R&D as well as management centers are organized to service a wide range of units. Since the critical resource is the experienced technician, physician, or manager, these individuals are grouped together and their activities carefully planned and allocated to serve the entire corporation.

Pfizer Field Operations report directly to two executive vice presidents and to the president. There are five regional groups: Pfizer-Asia, Pfizer-Africa and Middle East, Pfizer-Europe, Pfizer-Latin America, and Pfizer-Canada.

As can be seen in figure 7-1, the *directors* of Asia, Africa, Latin America, and the Middle East, report to the executive vice president; the director for Europe reports to the president, and Canada reports to the operations vice president. Each executive vice president and the president have area responsibilities as well as certain staff responsibilities.

TECHNOLOGY TRANSFER

Though the New York Headquarter Office is one of the major centers for technology planning and transfer, there are fifteen research and technical service centers established in the United States that focus on different products:

Brooklyn, N.Y.: Pharmaceutical, chemical;
Chicago, Ill.: Dental;
Easton,: Minerals, pigments, high temperature metals;
Greensboro, N.C.: Specialty chemicals;
Maywood, N.J.: Cancer diagnostics;
Milwaukee, Wis.: Diary and brewery products;
New York, N.Y.: Cosmetics, fragrances;
Old Bridge, N.J.: Refractories;
Parsippany, N.J.: Toiletries, consumer health;
Redmond, Wash.: Poultry breeding;
Rutherford, N.J.: Orthopedic products;
Terre Haute, Ind.: Agricultural;
Wallingford, Conn.: Metals.

Each of the above sites carries out specialized research, while overall basic research and development is centered in Groton, Connecticut, and Sandwich, England. Internationally, support research is carried out in Argentina, Australia, Belgium, Brazil, France, Germany, Japan, and India. Pfizer focuses on its regional management

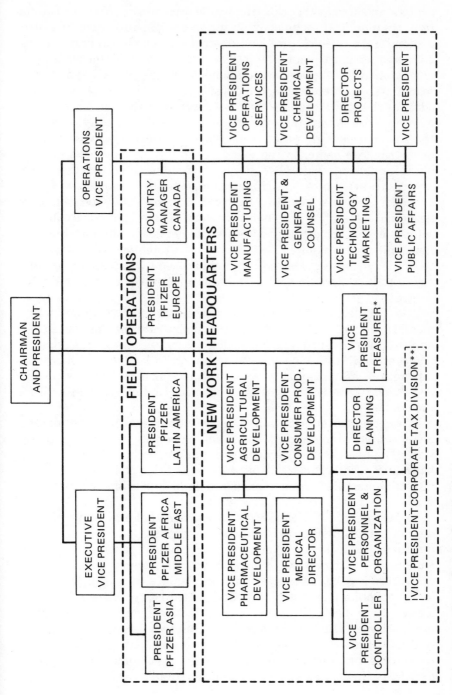

Figure 7-1. Pfizer General Organization

centers and its New York offices to centralize the flow of information, training, and technicians to support its global system. If a plant encounters a special problem or seeks research support, it first goes to its management center. The management center would provide immediate support with technicians and processes or would search its geographic area for consultants. If the problem requires additional support, the request is forwarded to Pfizer International in New York where the staff might deal with it directly or request the research centers, other management centers or any production or distribution center in the world to supply the required specialists or information. Naturally, Pfizer is remunerated for the direct services provided. At its foundation, the technology search and exchange system is keyed to full time staff specialists located at the headquarters and management centers. These individuals are responsible for being able to locate all necessary support that any field unit might require. They also provide essential expertise for the technology transfer. This same system is used to control and monitor new product development, production improvements and quality control.

The parallel activities for the specialized staff (pharmaceutical, medical, agricultural, chemical, and manufacturing) provide direct control over quality and specialized services, while the country manager and unit manager focus on profit and cost issues. The integration of high-quality scientific and medical goals with cost and market criteria is essential to balanced and effective growth for the company.

Pfizer spent over $10 million in 1974 on its international management centers alone and a similar amount on its New York support staff. The management centers' research and technical services alone have 264 technicians available to the global system. Overall R&D efforts represented more than 67 million dollars spent around the world.

In the long run, Pfizer's success will depend not only on its product developments but also on its effectiveness in introducing and modifying these products for a variety of cultural and geographic settings. Due to the complexity of local conditions, unique characteristics of culture and health patterns, the availability of raw materials, the local technology infrastructure, and skilled labor, Pfizer must adapt almost every new unit so that it can effectively meet peculiar market requirements. Though the long-range research is centralized in two centers (Sandwich and Groton), every dosage and bulk plant must be capable of providing its own product development and research to modify the basic knowledge for local requirements.

RESEARCH

At the heart of the pharmaceutical technology system is the discovery of new compounds and processes. Unlike academic research, Pfizer and other industrial research facilities normally attack the problem of discovery with interdisciplinary teams. These teams are formed within research centers that supply the most up-to-date support systems. The critical need for interaction at all levels of the research effort inhibits decentralization of "pure research."

Teams are normally made up of pharmacologists, chemists, biochemists, clinicians, and toxicologists. Each team is supported by a variety of technical services like data processing and research administration. Cross-fertilization of ideas is continuous and all teams are kept in constant communication with overall research plans. Any fruitful development can be followed up immediately. The centralized approach provides support at each step and allows maximum access to ideas and the most effective means of providing testing and field trials. Pfizer spends $67.7 million on R&D per year and employs 1,982 technicians and scientists.

On the average, a drug that finally reaches the market will represent an expenditure of $24 million, and 10 years will elapse between the initiation of research and final marketing.[a] Of a final selection of 100 projects that are nominated for major research, seven will reach the market. Each year the different representatives from divisions of the worldwide system meet to establish long-range research objectives. The goals aimed at are usually general areas of therapy like anti-infectives, cardiovascular, or gastrointestinal treatment.

Within the two major research centers, average support facilities for each scientist will cost in excess of $150,000 per year. His salary and personal expenses will exceed $45 thousand, two technicians to support him will cost the company $75,000, and even the less sophisticated laboratory will require over $80,000 worth of equipment for his particular use. At Pfizer in 1972 the following types of specialists were employed:

1. Medicinal Chemists (chemotherapy and noninfectious diseases);
2. Research Biologists:
   Psychopharmacologists
   Cardiopulmonary pharmacologists and physiologists

[a]Swartzman, David, *Characteristics of Drug Research and Research Activity and the Size of the Firm,* Department of Economics, New School for Social Research, 1972, (unpublished manuscript).

Gastrointestinal pharmacologists
Biochemists/biochemical pharmacologists
Immunologists
Microbiologists (bacteriologists, parasitologists, virologists)
Specialists in drug metabolism
Toxicologists (teratology, genetic toxicology, hematology)
Pathologists;
3. Pharmaceutical Chemists/Pharmacists;
4. Analytical Chemists;
5. Process Chemists;
6. Clinical Researchers (medical monitors-M.D.'s, statisticians, data processors, FDA liaison):
7. Data Coordinators;
8. Technical Literature Specialists;
9. Research Administration (planning and budgeting).

Most of these specialists have spent over fifteen years in one reasonably narrow area of drug development. The high degree of specialization is offset through the careful integration and planning of the multidisciplinary teams. Basic research is hard to decentralize. However, the laboratories at Sandwich and Groton do much more than regional R&D and just search for new compounds. They stimulate a wide variety of laboratory utilization in other Pfizer units.

In most instances, the main laboratories act as a source of information and guidance for problem solving in the field laboratories. When new tests are required or unique obstacles develop, the main research laboratories provide support or locate outside specialists to deal with the challenge. For example, the Groton facility was encouraged to create more effective bio-assay tests that could quickly and accurately determine the potency and sterility of certain drugs. The laboratory located a computerized scanning device and created programs that would allow electronic reading of test mediums for a number of biological assays. This technique and equipment is available to all laboratories as are all newly developed tests, procedures, and equipment. Basic research is complimented by the focus on creating new and efficient laboratory techniques.

The cost of the technology system at the headquarters and management center level is allocated according to sales and profit. Normally, the management center is allocated a portion of the corporate overhead and then in turn, allocates these expenses to different markets within its region. The company makes no effort to try to allocate the individual costs of technology support to each unit. The basic policy is that management must supply whatever technology necessary to develop a market; it is considered a cost of doing business. In an industry so highly dependent on innovation and specialized skills, it is necessary to coordinate different research, manufacturing, and training skills with a global approach for economies of scale. In addition, the rapidity of obsolescence also calls for a centralized[b] effort to seek out new developments and areas of product improvement.

Though the basic research is highly centralized, the drug development procedure in Pfizer stimulates technology transfer and R&D activities throughout the world. Since each product or compound must be modified to fit different markets, each drug must ultimately be adapted by local laboratories. In addition to this, the Pfizer systems are mobilized throughout the globe to carry out tests and experiments regarding different formulations either developed at the central laboratories or licensed from other sources. Any significant field unit is required to have its own research and development capability. These field laboratories must be capable of supporting manufacturing with sophisticated quality-control systems as well as developing technology suitable for managing animal and clinical testing. The field labs must also have the independent capability of modifying products, developing new compounds with local materials, and evolving the related procedures. Though basic research is not the mission of these field units, such capability must be transferred in order to carry out their normal manufacturing, product development, and testing activities.

The complex pattern of evaluation, trials, and government approval for drug acceptance places another extensive requirement on local R&D activities. At the minimum, all drugs must go through certain stages of testing and approval inside and outside of the company.

Evaluation within the Company:
    Initial chemical synthesis;
    Animal evaluation for efficacy (Does the compound work?);
    Animal evaluation for safety;
    Request approval for human tests.
Evaluation Outside the Company:
    Government reivews the initial research (Does it work and is it
        safe for further tests?);
    If approved, first tests in humans for absorption, duration, and
        toleration (clinical trials);
    Clinical validation (Does it work in humans, and did clinical
        trials confirm original evaluation of the compound?).

[b]Ibid., pp. 4-25B.

While the animal and clinical validations are going on, the company must also address the problems of how to manufacture, store, and package the compound. Both the questions of clinical validation and manufacturing technology development are tackled by different units of the corporation. To carry out these activities most field activities, especially those overseas, have developed their own R&D capability. As the local infrastructure grows to support more advanced activities, the overseas units gain more and more capability to carry out experimental research. At the moment many new formulations, dosage developments, and process breakthroughs come from Pfizer foreign plants and laboratories. An important prerequisite for developing the local infrastructure is the bringing together of local physicians, laboratories, and Pfizer professionals to carry out field testing and product adaptation. This type of technology transfer is often ignored as it is not linked to "pure research," yet it is a major method for stimulating the training and experience base for a research-oriented community of physicians, pharmacists, and chemists.

Within all of the larger overseas operations, much of the effort of the medical and veterinarian divisions is devoted to managing, testing, and validation programs that utilize local physicians, laboratories, and test centers. In many developing countries, Pfizer provides the initial training and guidance for these individuals to support their testing and validation needs.

PROBLEMS IN TRANSFERRING RESEARCH

In most areas, Pfizer would like to encourage the development of research at their production sites. The goodwill of such activities is sometimes coupled with lower operating costs through economies of scale in salaries and facilities. The local research activity also helps the company perform studies for drug registration and advanced product development.

Unfortunately, in many developing areas the obstacles to establishing laboratory facilities are substantial. Importation barriers often restrict access to critical equipment and supplies, and in many cases fixtures and equipment must be fabricated on the site. When importation of specialty chemicals and devices is allowed, customs and bureaucratic procedures sometimes limit importation due to administrative and management problems.

Most equipment must be purchased with hard currency, and this places further complications upon the already difficult purchasing and importing procedures. Since laboratory animals are usually not available, they may have to be bred and raised at the site. Finally, the lack of PhDs, physicians, technical typists, specialized information clerks, lab technicians, adequate power, refrigeration, and many other infrastructure necessities hamper an effective transfer of laboratory and research technology. Yet growth requires a scientific and technical base at every plant. The degree to which the research and development can be transferred is tied to the capability of the local infrastructure to support and receive the technology as well as the motivation of Pfizer to overcome the obstacles, thereby gaining access to emerging markets.

Many areas require specific product adaptation, and the more streamlined and simple registration laws of some developing countries has stimulated greater focus by Pfizer on these markets. Once a new compound or application has been developed, it can often times be introduced more quickly internationally. As a result, more of the long-term research of the company is oriented toward medications and treatments for developing markets. The local laboratories are called on to supply continuing information on new market characteristics that relate to flavors, color, size for the product, medical practices and local disease peculiarities.

EVOLUTION OF PLANT FACILITIES

Traditionally, Pfizer's first contact with a new market is established through an agent relationship. The local representative merely distributes and markets imported products. As the market grows, Pfizer may consider setting up a local distributing and dosage form of manufacturing facility. With increased demand, the company will take its first major step toward local production by constructing a dosage plant.

The dosage plant is a major step for the corporation as it requires a large committment both of technology and capital. The dosage plant must develop a technical capability for adapting bulk materials to local ingredients, materials, and applications. In one country, the compound may be administered by the use of a tablet taken orally, while another may require intramuscular injection. Though the bulk compound is the same, the complexities of sterile mixing, developing the right granulation, assuring proper adherence of powders, and other variations of production demand a sophisticated technical capability. Local materials, humidity, utilization of the drug, shelf life are all variables that must be dealt with at each dosage plant. The skills required for a basic dosage plant including those of compounding, quality control, production management, and production development are similar regardless of the size and diversity of the plant.

In the well-developed markets, Pfizer will ultimately add a bulk plant that ferments or synthesizes the basic compounds. These may then be used in local dosage plants or exported to nearby regions. The additional technology required for this type of activity relates to the particular type of bulk plant. Fermentation skills are required for antibiotic production, while other specialized techniques are related to the production of vaccines and so forth. The bulk plant is normally built upon the technical base established by skills developed in the dosage activities for combining and producing ointments, liquids, capsules, injectiles, tablets, and powders for human and animal use. A simplified model of plant evolution in the pharmaceutical and veterinarian products is shown in Figure 7-2.

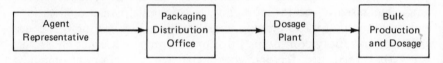

Figure 7-2. Simplified Model of Pfizer Plant Evaluation.

High risks result from attempts to compress the time this cycle normally requires. Some governments have purchased entire plants without developing infrastructure and basic technology skills. In Ghana, a bulk and dosage facility assembled by European consulting firms has been completed for several years but has not marketed its first products.

Pfizer has some form of dosage operation in 36 countries and has been able to add bulk production in ten countries besides the United States (Japan, France, Spain, Mexico, United Kingdom, Ireland, Brazil, Argentina, India, and Korea). In these ten cases, either the infrastructure and market could support a full line of activities, or the governments were ready to support the diseconomies resulting from beginning bulk production activities too early.

The first stage in the evolution of activities in a country is normally associated with the initial large scale marketing. At this point, Pfizer makes its first major technology transfer by developing the field sales force. This group of pharmacist-salesmen must be thoroughly trained in the drugs and must be able to communicate with physicians and local outlets concerning the elements of each disease that the specific drug is designed to treat. Pfizer documentation must be able to describe in detail both the characteristics of the disease and the Pfizer treatment, including the following elements:

Characteristics of the Disease:	Characteristics of the Pfizer Treatment:
Concept, history, definition	Origen and chemotherapy
Morphology, biological systems, and cycle of evolution	Pharmaceutical characteristics — absorption, metabolism, levels of effect
Clinical Forms and pathology	— Indications and treatment
Diagnostics	— Counter indications and precautions
Evaluation, prognosis, treatment	— Toxic effects and side effects
Possible prophylaxis	Resume of experimental data
Methods for evaluating new drugs and criteria for cures	Resume of country clinical tests
	Resume of international tests

The active sales force represents a vital link between new drug developments and the clinician who must utilize it. More than any other element, the technology exchanged through the field force (pharmacists) and the local physicians, hospitals, and pharmacies is the beginning of the full-scale technology transfer in the receiving country. With market growth and adequate scientific infrastructure, production technology will follow as soon as feasible.

During the early 1950s and 1960s, Pfizer and other drug companies were actively trying to stimulate dosage and bulk production in a wide variety of sites. Quality-control problems emerged and forced the companies to slow their transfers and maintain their centralized bulk production. By the late '60s stabilized bulk production was again being decentralized with Pfizer operating ten overseas plants. In some cases, Pfizer must pass up a promising market opportunity due to governmental demands for inappropriate technology. The threat of losing control of the quality or incurring extensive problems in establishing production facilities overrides the desire to penetrate the market. The consequences of a loss of quality control in any area of the world would jeopardize years of confidence and brand identity. The long-term policy of the company is based on quality, innovation, and cost-effective production.

Whether engaged only in dosage operation or the full compliment of dosage, bulk production, and other product development (feed supplements, vaccines, chemicals, and so forth), there are several critical elements that all Pfizer facilities must contain. To support all the basic production and test aspects the utility facility must be capable of generating steam, electricity, chilled water, salt water (brine), compressed air and sterile air with controlled relative humidity.

Effective and continuous provision of these different utilities is critical since a slight interruption in production can force destruction of the entire batch.

Quality-control and development laboratories are necessary for production modifications and provide extensive testing necessary prior to production runs (materials testing), during production (in-process tests), and after the production campaign is completed (finished goods tests). Additional laboratories are needed for a wide range of activities including extensive animal testing, organic chemistry and so forth.

Internal maintenance facilities must be capable of building new jigs, dies and control devices, as well as basic machining and repair. The shop must service everything from air conditioning to sterile equipment for glass, since local environments often lack these specialized support facilities.

Finally, the production itself is normally broken up into two major areas. The dosage form or pharmaceutical plant takes care of storage, dosage manufacture, and packaging the different products. The bulk plant is normally oriented to fermentation or synthesis of ingredients followed by purification of the desired bulk active ingredient. For the fermentation activities another special laboratory is required to concentrate on developing the proper mediums for the spore growth.

Dosage Plant Routines

The mainstay of Pfizer overseas activities centers around the dosage form plant. Different products require alternate types of design, equipment, and skills, but most plants ultimately must deal with the production of sterile liquids, nonsterile soluble powders, wet and dry tablet granulation, ointments, capsules, nonsterile liquids, and sterile powders.

The beginning of manufacture routing is in storage and warehousing. Careful control of environment and materials is maintained on a day-to-day basis. Each input must be carefully tested upon arrival and before its use in product manufacturing.

For sterile liquids and powders, the first step of production is the weighing and testing at the pharmacy. From there the flow moves into the sterile areas of the plant where solutions are prepared. Sterile filtration can be carried out by the use of a number of routines, the selection of which depends on local equipment and worker skills. Specialized ampule machines insure accurate filling and sealing.

For tablets and powders, many different steps of milling and blending may take place before the mixtures are wetted, granulated, dried, or tableted. Oftentimes, the tablets are coated and then have to go through specialized coating rooms and polishing pans. The operators involved in these production activities must be fully capable of operating turbulent and rapid mixers, granulators, a wide variety of tableting machines, vacuum dryers, polishing pans, coating pans, and powder and liquid filling devices.

Ointments demand the use of planetary mixers and roll mills to mix and homogenize the compounds. Specialized filling machines and testing devices are also required to insure proper tube filling and sealing. All these activities require sterilization, filtration, and cooling stages during the process. (See the typical flow chart shown in Figure 7-3.)

Whether producing powders, capsules, ointments or whatever type of dosage preparation, under controlled atmospheric conditions the manufacturing flow is interspersed with multiple quality tests. At Pfizer, quality-control personnel are stationed at all critical points of the manufacturing process. Examinations may be by hand inspection of tablets or specialized sampling of powders to be tested for pryogens in the quality-control laboratory. To insure appropriate authority in quality control, the responsible officer in this area may report directly to New York as well as to the plant manager. He has unquestioned authority to reject or approve any product in a plant until a quality review board can examine the situation.

Bulk Plant Routines

The major stages of bulk production in fermentation type drugs is spore production, fermentation, clarification, and recovery. A specialized quality-control lab, the fermentation-control lab, ensures that the fermentation process is developing the most potent batches of each basic pharmaceutical ingredient. Though the effectiveness of the spore is improved through research, production effectiveness can be easily affected if correct procedures are not followed strictly. A campaign for penicillin production requires at least 30 days, and any slight variation can cause the loss of the entire batch.

To establish the correct environment for growth and multiplication of the spore, a special center is established at the plant to mix and prepare the nutrient or media in which the spore will be multiplied. The media is mixed with the spore in the initial stages of fermentation. Specialized equipment is used to agitate, aerate, and control the growth of the organism. After three to seven days, the batch is moved into the clarification area where several procedures of filtration are used to remove the now dead organism from the

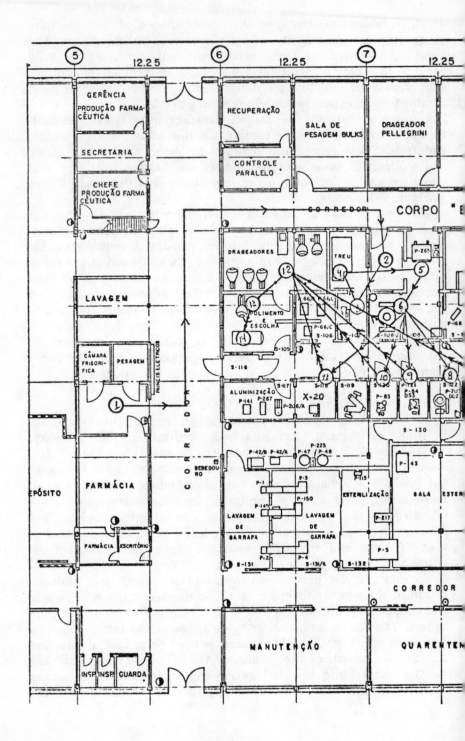

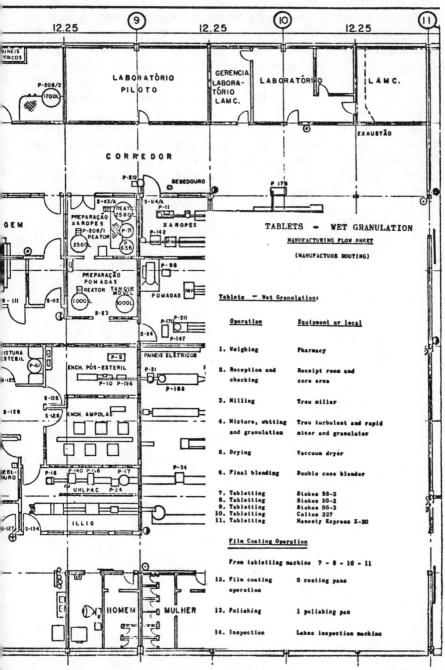

The following text appears within the manufacturing flow sheet portion of the figure:

TABLETS – WET GRANULATION

MANUFACTURING FLOW SHEET

(MANUFACTURE ROUTING)

Tablets – Wet Granulation:

Operation	Equipment or local
1. Weighing	Pharmacy
2. Reception and checking	Receipt room and core area
3. Milling	Trou miller
4. Mixture, wetting and granulation	Trou turbulent and rapid mixer and granulator
5. Drying	Vaccuum dryer
6. Final blending	Double cone blender
7. Tabletting	Stokes BB-2
8. Tabletting	Stokes DD-2
9. Tabletting	Stokes DS-3
10. Tabletting	Colton 227
11. Tabletting	Manesty Express X-20

Film Coating Operation

From tabletting machine 7 – 9 – 10 – 11

12. Film coating operation	5 coating pans
13. Polishing	1 polishing pan
14. Inspection	Lakso inspection machine

Figure 7-3. Pfizer Dosage Plant. Typical Manufacturing Flow Chart (Tablet-Wet Granulation)

effluent. Normally. it is the excretion of the organism that makes up the antibiotic or critical ingredient. Depending on the desired quality and purity, the broth might be kept here for several different filtrations and clarifications. In some areas of the world, special procedures must be developed since, for example, the more advanced rotary vacuum filters are not available or cannot be serviced. Pfizer makes major design and process changes for different areas depending on the raw materials available or the equipment that can be utilized.

At the final recovery stage other forms of filtration are employed to extract the ingredients of the bulk production. Throughout the bulk production, control personnel monitor all stages to protect the quality and efficiency of the production. After the recovery, the bulk product is transferred to the dosage plant.

Chapter 8

Pfizer–Nigeria

ESTABLISHING AN INITIAL DOSAGE PLANT

For Pfizer, the development of dosage manufacturing operations in Nigeria was a stage in a familiar pattern.

In many developing countries Pfizer had been aggressive in terms of early evolution of production activities. In Pakistan, India, and several other countries, Pfizer had been among the first international companies to establish full-scale dosage or bulk production, introduce sterile laboratories and production, and actively push for introduction of complex manufacturing activities.

The Nigerian operation began in the traditional fashion as an export market for Pfizer, handled through an agent relationship. By 1968, it was clear that the market was growing rapidly and its 80 million inhabitants would require a wide variety of health aids and chemotherapy medications. The government was also anxious to import technology and help establish an independent and self-sustained infrastructure as soon as possible. By 1968, the company had made commitments in three production areas, including pharmaceuticals. Joint ventures in production of livestock feeds and marketing of pharmaceutical goods had been established. A small plastics company that had also been formed handled certain packaging tasks. Though the scientific infrastructure and health care facilities in Nigeria were considered inadequate by normal standards (1,500 physicians for the entire country), Pfizer hoped to gain an early lead in the growing market and also demonstrate its commitment to local production and management.

Planning began in 1970 for the establishment of a full-scale dosage plant, to be operating by the fall of 1975. A well-developed system for planning and establishing initial programs for construction, start-up, and value engineering placed initial emphasis on market surveys and production requirements. The investment of time and people in the early stages provided control, guidance, and operating plans for the successful design and adaptation of the Pfizer capabilities for Nigeria.

PLANNING AND BUDGETING TECHNOLOGIES

Over the past years, Pfizer has evolved a specialized approach to planning and controlling the construction and start-up of new plant facilities. Headquarters demands a series of related control reports and supplemental plans to support critical assignments and decisions during plant development. The required management tools aid the company in bringing all its global capability to bear on the technology transfer required to launch the new plant.

For each new plant, a project manager will be assigned and will normally report to one of three management groups (New York international headquarters, the regional management center, and the country management center). He will coordinate with all three. Both line and staff support will be assigned from the overall company to the new project.

Though the proposal for a new project may come from any source within Pfizer, as can be seen in Figure 8-1, it will be assigned for further analysis to the particular country manager. From the country manager, initial marketing and strategy information will make up a "probable Project Proposal." The manufacturing division of the management center will provide the initial technical information suitable for preparation of the "Preliminary Design Basis." Experienced chemical, production, and industrial engineers evaluate the production requirements, develop process descriptions, flow sheets, and equipment layouts. To ensure an appropriate proposal package, two engineers visited the African management center and Nigeria for five weeks to assemble the initial data. During an earlier visit by this same project team a suitable site was selected and an architect appointed. The availability of this planning expertise helps ensure that reasonable projects are not frustrated by inadequate planning and lack of local technical skills.

If the market and infrastructure look reasonable, the technical research will continue to determine sources of raw material and

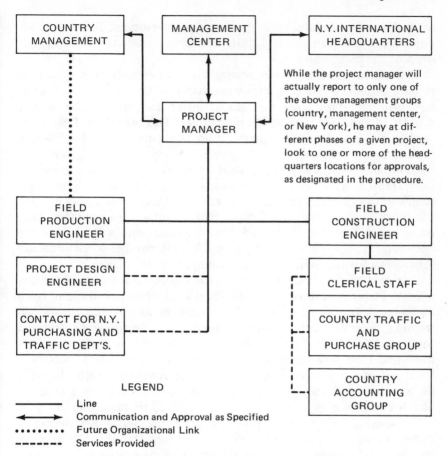

Figure 8-1. Pfizer Project Organization (Functional Relationships)

prepare equipment requirement lists and manufacturing costs. With 78 foreign plants successfully completed in 37 countries, Pfizer is capable of supporting new plant evaluation and planning with a wide variety of technicians and accumulated experience. These planning and management skills represent a unique technology often overlooked since they precede any actual negotiation or commitment, yet are the basis for successful future operations.

From the estimates incorporated in the "Preliminary Design Basis," the country manager and his staff can outline the expected steps for managing the development of the new plant. In addition, the overall plan (P-3 in Principle) can be submitted to New York headquarters for review and modifications prior to final programming. Depending on the complexity of the host country, environment, and legislation as well as the availability of professional personnel, Pfizer will appoint a project manager and designate the appropriate line and staff support for his mission.

The duties of the project manager are essentially that of overall project coordination and implementation of the project plan. Construction supervision and equipment installation will be the responsibility of the field construction engineer. The other key field manager is the production engineer, who concentrates on establishing production procedures and techniques. The production engineer may become the plant manager at start-up until a local manager can be fully trained, which normally requires at least two years. During the stages of design, construction, and start-up, these three Pfizer specialists work closely with the future plant engineer, maintenance foreman, utility foremen, and workshop foreman, who will become the backbone of the plant operating staff. This parallel work situation is the basis for the major transfer of management skills.

During the first two to three years of construction and start-up, the Pfizer specialists and their local counterparts will spend from 40 percent to 66 percent of their time in formal and on-the-job training. For the initial Nigerian technical start-up work force, a total of 1,000 man-days of training will be provided. As the plant develops, new personnel must be added, and individuals who will leave for other jobs or positions must be replaced, and both must be adequately trained. Pfizer estimates indicate that in a country like Nigeria, it may lose up to 20 percent of its managers per year to competing industries or government. Table 8-2 shows the focus Pfizer places on experience as well as in-plant training. Considering the emphasis placed on quality control and in-depth manufacturing know-how, the Pfizer trained individual becomes an important resource to the country and to other industries who may later hire him.

Before final approval of the Definitive Plant Development Plan (see Table 8-1), several detailed studies and technical plans are developed. The project design engineer is designated at New York headquarters to support the project in all design activities. To complement the basic production and facilities design work, other technical studies will be carried out with the help of the management center (water and soil surveys, site selection reports, and so forth). To coordinate supplies and global purchasing, a team from New York headquarters of the management center staff is designated for the life of the project. Availability of items such as cement, steel, cinder blocks, and piping will radically alter the type of design and construction plan finally proposed. For Nigeria, building materials

Table 8-1.   Pfizer–Nigeria Plans and Assignments during Planning, Design & Construction, and Start-up Stages, 1974

	PLANNING & PROPOSAL STAGE	*DESIGN & CONSTRUCTION STAGE*	*START-UP STAGE*
MAJOR CONTROL REPORTS	PROBABLE PROJECT PROPOSAL PRELIMINARY DESIGN BASIS PROJECT MANAGEMENT PLAN DETAIL PLAN (P-3) IN PRINCIPLE	PREPARE DEFINITIVE PLOT PLAN DEFINITIVE PLANT DEVELOPMENT PLAN (P-3) ARCHITECTURAL DRAWINGS CIVIL CONSTRUCTION SPECIFICATIONS	PROJECT STATUS REPORTS PROJECT CRITIQUE REVIEW OF PROJECT CRITIQUE
SUPPORTING REPORTS & PLANS	Provide Basic Technical Information Provide Basic Marketing Information	Design Schedule Site Selection Report Equipment Lists Equipment Layout Site Work Design Process Equipment Drawings Utilities Drawing and Specifications Air Conditioning Drawings and Specifications Laboratory Equipment Drawings and Specifications Electrical Equipment Drawings and Specifications Report on Import, License and Duty Regulations	Procedures for Local Purchase Procedures for Cost Control Purchase orders Equipment Inspection Guides Tests for Raw Materials Operating Manuals
DECISIONS AND ASSIGNMENTS	Appoint Field Construction Engineer Appoint Field Production Engineer Designate Project Design Engineer	Site Selection Designate Contact for Global Purchasing and Traffic Select Pfizer Architectural Specifications Issue Drawing Lists Select Local Architect Accept Bids for Construction Accept Bids on Equipment	Form Local Project Staff Secure Import Licenses Hire Plant Staff Train Plant Staff

were not a problem in comparison with the equipment of which almost 100 percent had to be sourced from Europe and the United States. For many of these items, lead times of twelve months were required and careful planning was demanded to ensure that equipment and facility preparation were coordinated.

The newly designated supply staff cooperates with the line engineers and project manager to work up equipment layouts, site designs, and a wide variety of equipment drawings. Eight to thirty different specialists may be involved in these early planning and design activities at New York, the management center, or at the plant site itself. Though local construction and architectural firms are always used, the specialized characteristics of pharmaceutical plant design must be transferred to the local firms. Using carefully developed guides, drawings, and manuals, the Pfizer engineers will spend some time with the local firms helping improve their expertise in the special problems of pharmaceutical plant design. Through contact with these individuals and access to the corporate specifications and other materials, Nigerian construction and architectural firms gained new expertise suitable for all types of sterile and controlled installations, food processing activities, and other chemical plant construction.

The various reports and studies validate the guidelines and goals of the formal plan, which, upon approval, clears the way for field staff and support personnel to begin their negotiations and contracting with the local suppliers and construction support. Unique problems may require specialized aid from corporate headquarters. In Nigeria, the problem of conflicting tribal claims to land makes the question of securing clear title for construction a serious issue. Pfizer attorneys drawing on prior experience helped solve this question after spending several man-months on the issue.

Once the plant program is underway, Pfizer calls a series of critiques and reviews of the project to accumulate comments and lessons applicable to the next situation. Manuals and guides will be reworked and new material added to the files as technicians gain experience. In Nigeria, Pfizer is the first to develop a complete dosage plant with appropriate sterile production capability. Coupled with its technology organization, Pfizer can bring the right type if expertise and problem solving skills to bear on the unforeseen problems.

DESIGN AND CONSTRUCTION TECHNOLOGIES

Pharmaceutical manufacturing requires an extremely reliable production area resulting from strict construction specifications. Sterility of storage and work areas demand construction standards that had not been previously required in the country. Neither the basic architectural design technologies nor the actual construction technology existed to the degree required by Pfizer. Headquarters and the African management center utilized a wide variety of written materials as well as specialized technicians to guide the work. The exposure of local workers, engineers, architects, and managers to these documents and individuals represents a considerable transfer of technology.

The principal technology transfer for the design and construction stage came through the field construction engineer. The Nigerian plant required both an experienced construction and production engineer to work with the project manager for four years. The job of the construction engineer is to train and develop the local design, construction, and future plant maintenance personnel. The field construction engineer came to Nigeria after successfully managing the development of a similar plant in Indonesia. Previously, he had spent twelve years on different construction sites for pharmaceutical plants. His specialized skill, building dosage plants in developing areas, is probably available in not more than twenty men in the world, normally located with large transnational corporations specializing in international pharmaceutical production.

The field construction engineer for Nigeria arrived soon after final agreement on the land had been arranged and New York headquarters had approved the definitive plan of action. At the same time, a national counterpart was hired to work with the field construction engineer on all stages of the development. The training and exposure of the local engineer during the two to three year construction and start-up would provide a basic knowledge of the plant necessary to take over the major engineering functions. Locating a competent engineer with some experience in any aspect of industrial manufacturing was a serious problem. There was no attempt to find someone skilled in pharmaceutical engineering. Pfizer prefers related experience but not extensive exposure to pharmaceuticals so that they can train him in their own specific approach to plant engineering.

To support the initial architectural drawings, New York supplied a series of concept and detail drawings. If the industrial experience of available firms is not adequate, Pfizer architects and engineers will visit and work with the firm. In Nigeria, this was accomplished with the support of the field construction. The scope drawings give room size and arrangement, but leave flexibility for adaptation to local materials. Sometimes materials like steel cannot be easily utilized due to the inexperience of local contractors. The field construction engineer helps work out the best combination of materials to meet the construction specifications and match the capabilities of the contractor and the available local materials.

With a basic agreement on the construction design, the field construction engineer will work with his counterpart to outline the basic engineering requirements of the plant. New York headquarters normally supplies the specific engineering drawings that will cover mechanical engineering, electrical engineering, and air conditioning requirements of the plant. The manufacturing and engineering staff in New York may make extensive recommendations and pose alternatives to the field construction engineer that are based on the accumulated experience of the support staff.

In Nigeria, few contractors had ever been forced to work with detailed air-conditioning drawings. Normally they would just adapt the commercial comfort air-conditioning equipment in an ad hoc manner. Since this is unacceptable for a pharmaceutical manufacturing plant, the capability of two supporting construction firms to develop specific air-conditioning plans and drawings was improved. For the plant facility, 40 to 50 main drawings are required from New York along with 10 to 15 thousand specification sheets and equipment drawings. From these documents the local engineering staff assembled the specification manual and equipment manuals for the plant. These documents form the basis for working with the subcontractors and suppliers for finalizing bids on planning. Many of the suppliers were given technology assistance so they could meet the tolerances and specifications of the plant.

As the different design drawings and architectural plans were adapted, construction reviews were carried out with the government. Pfizer had learned in Indonesia and other countries that effective contact with the government was necessary not only to get approval of designs, but also to help the government technicians understand the new procedures and equipment that would be utilized. In Nigeria, many meetings were held to explain the new methods of welding stainless steel, installing and operating pressurized vessels, and the pressure tests for the new vessels. Sometimes, such contact encourages the government to plan for creating other industries that can utilize the initial technology imported for the dosage plant. Manufacturing the stainless steel equipment locally has become a possibility because of the availability of Pfizer's testing and specifications for their stainless steel equipment and facility components.

As the actual plant construction begins, the field construction engineer assumes the daily activity of monitoring the work in progress and suggesting improvements or modifications. Much significant technology is transferred to the architects, construction managers, and workers of the contractors and subcontractors in this way.

In Nigeria, almost no effort has been put into training the lower level laborers. Most of these workers were temporary, brought on only for the construction phase. Pfizer engineers spent a great deal of time seeking more effective training for these individuals to keep the work process on schedule. For example, the field construction engineer developed a tool box list and instructions to guarantee that each worker would have the right tools available and know how to use them. In addition, a tool box was designed on wheels so that the worker could pull the box around with him, rather than leaving the heavy box in one area and bringing his tools back and forth.

Basic construction fixtures often constitute a type of technology that may be new to the local environment. At the new plant, the Nigerians were planning to erect a roller shutter door by placing the setting in holes in the wall, then filling them with cement. The field construction engineer knew that this antiquated approach was often highly inaccurate and caused extensive door problems. He suggested a new method of placing a "T bar" on the frame with expansion locks. With the frame in place, holes can be drilled that are easier to measure and give a better fix to the wall. This method had not been utilized by the construction workers before.

Concrete drills were replaced with more modern tools. In the past, the contractors did not want to train the workers in sharpening the bits so they had accepted the less effective drills. Expandible plastic anchors were also introduced by Pfizer as a better way of fixing all types of devices to walls and floors. These simple devices had never been imported into Nigeria before.

For wiring, the Nigerians had often simply used matches to stuff the wires into place. The field construction engineer demanded proper wiring anchors and also introduced the workers to compression type couplings that easily allowed connections for conduit. No threading was required and the system could use a lighter and more durable pipe. These simple techniques, along with training the workers how to use the right tools, resulted in better efficiency in plant construction and also provided new capabilities to the Nigerian contractor and workers.

As the steam generators are installed in the next stages, the workers will have to learn how to operate and maintain the proper types of piping and valves. Some of them will be trained in how to monitor the gauges and recorders. The focus of Pfizer on the most effective construction will require many new skills to manipulate the "steam flow recorder" and other devices critical to good manufacturing control and management. To ensure a facility that meets specifications, the company and the field construction engineer must spend time and money to educate and develop the capabilities of the designers, architects, construction managers, and even the routine workers.

START-UP TECHNOLOGY

With the basic designs, layouts, and processes transferred, the development of the plant shifts from active construction of facilities to the development of the manpower and skills necessary to actually run the facility. From the management base established previously through the marketing activity, Pfizer has to create a start-up manufacturing staff and develop the quality control systems, industrial engineering production planning, inventory control and other routine procedures for a stable manufacturing activity. In contrast to the strong relation with the outside environment for construction, architectural and supply information during planning, design, and construction, the start-up period witnesses the most dramatic internal transfer of standard manufacturing technology for the dosage activity.

For full-scale dosage activities, the plant must train and manage four classes of employees. First, management and professional levels skills must include mechanical engineering, finance, chemistry, medicine, industrial pharmaceutical engineering, and electrical engineering.

Second level management includes all foremen while "weekly personnel" normally refers to the clerical and technical helpers. The fourth category is made up of hourly workers.

The expansion of marketing and distributing efforts in Nigeria led to the development of the general administrative and marketing staffs suitable for later expansion. Within the Nigerian region, Pfizer had established a medical advisor and staff to help prepare the detail men (medical salesmen). The overall administrative group included a controller's department and a personnel management group. The only active field groups were in the agricultural department, which distributed animal health products and also helped finance and manage joint ventures throughout the country for livestock feed mixing and sales. When the pharmaceutical plant was approved in 1972, a field construction engineer was assigned to Nigeria along with an experienced production engineer, as plant manager. These two individuals began forming the nucleus of the new plant management and skilled labor force. The first manpower training objective was to secure and begin training the national plant engineer, quality-control manager, production manager, PPIC manager (production planning, inventory, and control), and packaging manager. This nucleus would receive the most extensive training at European and American facilities. They, in turn, would prepare the production, maintenance utility, and workshop foremen. The technicians and workers would be hired after the first twelve months of training for first-line management had been completed. The plant engineering and maintenance group alone would require 1,751 mandays of training for fifteen men and possibly on-going training to replace turnover of 25 percent a year (see Table 8-2).

When fully operational, the plant would require production departments for sterile packaging, oral packaging, liquid and powder manufacturing, and tablet and capsule manufacturing. Support staff would have to be trained in materials management, value engineering, warehousing, quality control and systems engineering. For the fully operating pharmaceutical production activity, the organization might resemble that shown in Figure 8-2.

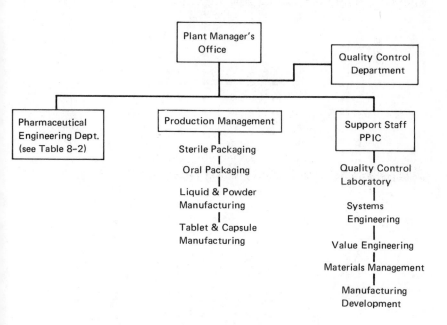

Figure 8-2. Pfizer-Nigeria Production Activity.

Table 8-2. Pfizer-Nigeria Start-up Training, (1975-76 (Plant Engineering Department Personnel Only)

POSITION	NUMBER	EXPERIENCE REQUIREMENT		TRAINING		Travel and Exchanges	FUTURE REPLACEMENT TRAINING @ .20 per Year
		Exp.	Formal	Pfizer OJT	On-going		
Plant Engineer	1	10	BS	218	20	10	50
Maintenance							
Foreman	1	10	HS	290	10	5	61
Mechanics	3	2	HS	290			174
Electrician	1	5	HS	145			29
Instrument							
Mechanic	1	5	HS	145			29
Utility Foreman	1	5	HS	88	10		17
Utility Attendants	3	1	HS	88			51
Workshop Foreman	1	10	HS	290	10		60
Lather Operator	1	5	HS	44			9
Carpenter	1	5	HS	44			9
Painter	1	1	HS	44			9
TOTAL	15						

Note: 1 man-year = 220 work days.

To carry out the different activities in engineering, production and PPIC the plant will initially require 138 managerial and direct labor personnel. The largest part of the plant personnel (51 percent) will be involved in plant services such as laboratory, engineering, warehousing, and so forth. Another 42 percent will make up the manufacturing services, with the remaining 7 percent involved in plant administration. Table 8-3 describes the different types of manpower allocation within the plant.

By the fall of 1975, Pfizer would have completed hiring for the 138 different positions with on-the-job training progressing while the first runs for producing finished packs of terramycin capsules were being initiated. By the end of 1975, the plant will be in operation, working up to an annual rate of 10 million in 1976.

Locating and training the production manager was the first objective of the management manpower program. Besides the extensive on-the-job training this individual received from the construction engineer and production engineer, he would be exposed to a wide variety of Pfizer activities in Europe and the United States. Pfizer management relies heavily on the direct exposure technique of training (OJT), complimented with specific formal courses. Though the local production manager would come to the company with an extensive professional background, the direct training at Pfizer facilities would be the most critical element of his technical preparation to manage the plant's production activity.

Pfizer selected a Nigerian who had gained a B.A. in Pharmacy in Nigeria and then received a Masters of Science and a Ph.D. in pharmacology in the United States. After lecturing at the Faculty of Pharmacy at the University of Ife in Nigeria, the new manager joined the company in 1974 to begin working with the engineers and project manager.

Outside of the direct experience transferred by the assigned specialists, the new production manager was sent to Brussels to observe and train on a variety of Pfizer processes. Major emphasis of the training program is always the relation between production and quality-control procedures. Other general areas include production ordering and recording, weighing procedures, production of nonsterile tablets, capsules, and powders. In the Nigerian case, the existing emphasis on veterinary products required time spent observing and working with blending and filling operations for veterinary powders. For all of the procedures expected to be installed in Nigeria, the manager would evaluate slugging and granulation techniques, compression operations, encapsulation, blending, filling and a wide variety of techniques used in various packaging, blending,

Table 8-3.   Pfizer-Nigeria Estimated Dosage Plant Manpower Distribution

	ADMINIS-TRATION	SUPER-VISION SCIENTIFIC TECHNICAL	ACCOUNT-ING CLERICAL	DIRECT LABOR	REPAIR-MAINT. LABOR	OTHER LABOR	TOTAL PERMA-NENT STAFF
PLANT ADMINISTRATION							
General Plant Administration	1	—	1	—	—	1	3
Production Management	1	1	1	1	—	—	2
Plant Accounting	1	—	3	—	—	—	4
Total Plant Admin.	3	1	5	0	0	1	9
PLANT SERVICES							
Engineering Administration	1	—	—	—	8	—	10
Repair and Maintenance	1	—	—	—	—	2	2
Utility Services	1	—	—	—	—	5	10
Laboratory Services	1	3	1	—	—	—	6
Internal Plant Warehousing	1	2	—	—	—	5	16
Finished Goods Warehousing	1	2	—	—	—	15	3
Purchasing	1	1	2	—	—	1	4
Production Control	1	1	—	—	—	—	1
Personnel	—	1	—	—	—	—	4
Cafeteria	—	—	—	—	—	4	—
Site Housekeeping and Security	1	—	1	—	—	17	19
Total Plant Services	7	10	4	0	8	49	75
MANUFACTURING SERVICES							
Basic Product Manufacture	—	—	—	22	—	—	22
Finished Product Manufacture	—	1	—	45	—	4	50
Packaging	—	—	—	—	—	—	—
Other Manufacture	—	—	—	—	—	—	—
Total Mfg. Services	0	1	0	67	0	4	72
TOTAL PLANT MANNING	10	12	9	67	8	54	156
Temporary Staff							

and milling activities. With Pfizer contemplating the first full-scale sterile production area for Nigeria, training would also include elements of sterile drug manufacturing, such as sterilization, setting up sterile areas, washing vials, dry fills and the activities of monitoring work in process and packaging within the sterile area.

To accomplish all of these activities Pfizer kept the new manager overseas for five months, (two months in Brussels, and three months in the United Kingdom.) Training for the quality-control manager would encompass a total of two months in Brussels and four in the United Kingdom.

For quality control, a Nigerian was selected who had a formal education including a Masters in Science in the United States and his Bachelor in Science (Chemistry) and Ph.D. (Chemistry) from Nigeria. As before, the experience prior to joining Pfizer was in the University of Ife as a lecturer.

The importance of quality control to the company is reinforced in many ways, including extensive training of the management. Laboratory organization, use of main instruments, and work in chemical analysis (standards, specifications, sampling, production orders) are primary to the initial training. Some exposure is provided for the routine manufacturing operations but more emphasis is placed on managing chemical analysis and observing the tests on work in process, raw materials, and finished goods. For all the products to be produced the trainee will study Pfizer techniques for carrying out bio-assays, stability studies, and other tests for the varieties of tablets, capsules, liquids and sterile products. Veterinary quality control is also an element of the training program.

Upon return of these individuals, full-scale training of foremen and workers will begin with their assistance. Pfizer maintains the presence of the project management specialists to ensure continuous interaction on production and quality-control activities for at least a year after initial production begins or until the Nigerian staff has acquired sufficient management skills.

Already assembled for the new management is a complete library on the general test procedures, compounding procedures and specifications, and finished product specifications. In addition to the procedures and processes supplied for the initial plant engineering, the parent company provides indexes and basic references on 3,010 different formulas and manufacturing instructions. The laboratory and management group can reach beyond their own library of manuals and guides to request information directly from their management center, the New York headquarters, or any other foreign plant on production problems, ingredient specifications, materials, formulae, or test procedures. At the Nigeria plant, the new production library includes detailed specification on compounding techniques, tests, and specific instructions for product manufacture. These manuals will be thoroughly utilized during the introductory period with the aid of temporary experts to ensure long-range growth and ultimate independence of the production facility.

Production and quality control are not the only areas that receive extensive documentation and training support. In the controller's department, the director was sent on several trips abroad for familiarization with procedures and training. Two weeks training assignments to both Sandwich and the management center in Nairobi developed an understanding of Pfizer accounting techniques. An additional one-month management training program in Sandwich was coupled with visits to other regional centers for exchange of ideas and problem solving. The manager claims that the training and the back-up procedural manuals act as his "daily bible" while the monthly reports and flyers from other companies in his area provide a global contact with other professionals operating with similar problems. The manager and regional controller will travel yearly to the Nairobi management center and other European plants to continue their training. The plant engineer, presently training with the project engineer, and the maintenance engineers will receive this same exposure and access in their own areas of expertise.

For Pfizer, management training is an essential element of manpower development, but a costly one. Due to the large demand for experienced personnel outside the company, some specialists leave after their initial training and exposure without making any actual contribution. Recently, Pfizer lost their first Nigerian quality-control manager after he had undergone extensive training for nine months in the United States and Belgium. Each time an individual is sent out of the country for training, the cost of his absence covers not only salary and air fare but fees for daily expenses ($50 to $70 per day), special clothing allowances ($250 to $350), specialized secretarial support, and loans for car and housing. These benefits must often be provided on a regular basis to demonstrate commitment by the company to improved capability for the individual.

The Pfizer Nigeria operations are required by government regulation to train and develop local management. Though it is a corporate policy to develop local management as fast as possible, (India, Korea, Pakistan are just three of the many foreign plants developed within the last four years with 100 percent local management and labor.) The Nigerian government has placed a specific quota system on all companies to accelerate nationalization of management. Unfortunately, this brings with it certain problems, while placing pressure on the training and development of local management.

Within most of its operations, the company has been able to successfully maintain its standards through hiring and training competent Nigerians. However, the inability to hire foreign technicians hampers OJT that might be improved if the option were available. The need for technical support for training is best seen in the experience of Pfizer in its plastics operation. This activity was prematurely restricted due to the inability to bring in foreign managers. With no qualified locals or technical expertise available in plastics manufacturing, the subsidiary has had to reduce its scope. Since Pfizer has lost its quota for foreign management in this plant, it has reduced its expansion, which could have meant the introduction to Nigeria of a wide variety of plastics forming and moulding technologies.

New York headquarters and the African management center are actively involved in all the various training and manpower activities involved in the start-up phase. Manuals, technicians, and visits are arranged to evaluate the new skills as quickly as possible. Pfizer has even developed a wide variety of teaching aids including the "teaching carroll" which is a sophisticated training system for presenting to the detail men (pharmacist-salesman) a wide variety of topics including chemistry, pharmaceutical nomenclature, and the active characteristics of drugs and diseases. Since the detail man is the link with the clinician and pharmacy market, it is critical that he understand and be capable of discussing the basic medical information and technology of the firm.

The New York technical staff is also called in for unexpected problems that emerge at start-up. As in the case of the South African plant, Pfizer engineers may have to develop new tubing and equipment to control local corrosive water characteristics. Other specific aid comes in the form of helping locate scarce supplies or devices that cannot be purchased within Nigeria. In some cases in Nigeria, Pfizer has made available financing to local suppliers to enable them to retool or purchase new machinery to supply the finished plant. Steam energy control was a start-up problem in Nigeria that required New York engineering assistance to locate a new type of steam flow meter for adaptation to the unique characteristics of the Nigerian plant. The headquarter's staff helped draw up specifications and procedures for utilizing the new meter.

When initial production begins, pharmaceutical manufacturing engineers and quality-control personnel may be dispatched from Nairobi and New York to help stabilize the first runs. These additional people help provide back-up for local technicians and additional training for the Nigerians. Foreigners will be assigned for extended stays in other areas such as cost accounting and specialized quality control to complete the initial start-up training. The nucleus of manpower will be expanded and all personnel will continue to receive technical materials in written form as well as the routine visits from the foreign centers.

The Nigerian staff will expend $50,000 to $70,000 a year in travel expenses alone to visit the Nairobi Management Center for on-going training as well as technical visits to Pfizer plants in Europe and the United States. Attendance is also required at global conferences on pharmaceutical manufacturing, quality control, personnel, and other technical areas as part of the continuing technology transfer that assures the continued evolution of the plant. At a minimum, the plant engineer, production manager, quality-control manager and manager of PPIC will make four visits outside Nigeria for training and exchange of information on problems and new techniques.

Since few companies locally stress process and quality control to the degree Pfizer does, the local plant will spend up to 70 man-days exposing workers, lab technicians, and other hourly and weekly personnel to their direct tasks and to in-depth work with the Pfizer system of manufacturing and control. Normally, the workers only have rudimentary education and possibly some exposure to compounding skills used in food processing or basic chemical production. For many, the training will begin with how to use rulers and simple scientific tools and then progress to the actual maintenance and operation of the blenders, encapsulators, slugging machines, and so forth. The laboratory workers will receive even more extensive training in the use of equipment and techniques for laboratory procedures. Throughout the training, Pfizer will emphasize "attitudes for quality control," which experience has proven is more essential than even the task skills they will learn.

To create the skills necessary to support production and full-scale laboratory testing, the company combines both long-term training and education experience with direct exposure through on-the-job training. The present production manager compliments the construction engineer with over fourteen years experience in Pfizer and six years of other chemical and pharmaceutical work. His background and continuous improvement in skills for production activities, along with the other foreign engineers, form the cornerstone of start-up training. Once the core Nigerian management is prepared and the first year of operation is completed, the plant will be capable of pursuing refinements and adaptations for the best operating procedures for Nigeria. With the help of outside specialists the Nigerian staff will have orchestrated eight major activities during the year and a half of start-up activities:

Start-Up Activities:	*Time Period Involved:*
Bulk forecast ordering and testing	6 Months
Ordering, testing, compounding ingredients	7 Months
Ordering, importing and testing packaging materials	12 Months
Locating and testing local packaging materials	7 Months
Training first and second line management	12 Months
Training production workers	2 Months
Creating plant procedures	6 Months
Start-up tests	1 Month

VALUE ENGINEERING TECHNOLOGY

After a plant has stabilized its initial production activities, efforts are placed on seeking new ways of reducing costs, improving production quality, and adapting the basic system to improve efficiency. The Nigerian staff, having been trained and guided through design, construction, and start-up will be capable of taking a more direct role in the on-going plant evolution. To support that evolution, routine exchanges, documents support, and exchanges take on a more expansive focus. For Pfizer, the regional management center plays an important part in the on-going technology transfer and exchange.

Role of Management Center

The African Management Center, which directly supports the Nigerian expansion, is organized by function with directors and staff in agricultural development, medical affairs, pharmaceutical development, manufacturing, personnel, and finance. With 18 professionals and 23 technical and support staff the management center acts as a concentrated technology pool for all African units. For manufacturing, the center provides support of production, process, and product development. Quality control, engineering, and construction, and indirect services (purchasing, traffic, planning, inventory) are also functions of the management center.

Periodic audits of country operations are an important factor for determining new means of increasing efficiency, cost performance, and implementing new procedures. These audits are coupled with the special assistance and trouble-shooting support that the specialists provide the Nigerian operation.

On-going information exchange is also facilitated by periodic manufacturing reports to New York and to the different African units. The management center also provides general technical information that is applicable to specific plants and environments. New manufacturing procedures or process changes developed in New York are reviewed by the center for application in Africa.

Quality control is improved and assured by the distribution of documents containing information from other quality-control laboratories, stability data, chemical products specifications, and new analytical methods or equipment designs. Technical assistance and quality-control audits are designed to keep the center in contact with the plant and ensure vigorous quality control.

Engineering for new projects, new programs, or modification is provided by the management center. Strong support in production engineering was essential to the development of the Nigerian plant, as was the center's aid in site selection, plot planning, and development of basic flow sheets and diagrams.

Purchasing, traffic, production planning, and inventory control are elements of general manufacturing services, which the center tries to accentuate and improve. These general activities might be supported through audits, technical assistance, or more directly through manpower training. Whether a new project or a plant expansion, the management center is designed to maintain specialized up-to-date resources such as publications and technicians.

The management center acts not only to provide on-going technical assistance but to establish an effective regional control of Pfizer activities and develop manpower for future expansion of the company. Individual units are not encouraged to invest independently in long-term development of personnel or services. The management center acts as the long-range focus in the region and also provides a convenient means for gathering information on new markets and developments that can be exploited by the individual units for future expansion. The management center de-emphasizes ad hoc development and centralizes it under a standard corporate growth policy. For example, in Africa, the push for indigenization of management could threaten performance standards and quality of Pfizer activities if not dealt with through a systematic manpower plan.

Management center training is designed to improve both the individual employee as well as the technical group in which he operates. The overall strategy of training is based on four major types of programs:

1. On-the-job learning supported by guidance from the country management center, or headquarters expertise;
2. Working/learning visits to other countries or technology centers;
3. Developmental job assignments of six months to three years inside the country at other units;
4. Formal training programs conducted inside the company.

The formal training programs normally focus on both general management and functional skills. During 1975, the African Management Center will provide formal programs supported both by New York and by their own offices.

New York-Sponsored Programs:
    Finance for nonfinancial managers (35 Managers, five locations);
    General management program (60 Managers, major markets).
Functional Skill Programs:
    Marketing management (all field unit marketing depts.);
    Accounting principles (30 clerks);
    Leadership of salesman (continuous for detail men);
    Production planning and inventory control (20 managers in four regions);
    Sales training (200 detail men).
Individually Tailored Programs:
    Department managers (production and quality control);
    Chemical management;
    Regional management training.

The most extensive and highly developed training is aimed at the continuous preparation of the detail men (pharmacist-salesmen). These registered pharmacists must be constantly retrained in the characteristics of the drugs and diseases of the plant and region. For the physician and pharmacist, the detail men of the different pharmaceutical companies are the major link to new drug and treatment developments. Pfizer must be able to represent its products and their use clearly and effectively to gain increased use by the medical community. In a country like Nigeria, with only 1,500 physicians, the role of the detail man is critical in terms of keeping the clinician informed and reducing his time commitment for seeking outside information. The medical knowledge transferred through these sales representatives is a substantial technology transfer that benefits the overall health services of the country. Over 30 percent of the management centers' efforts are devoted to bringing new training aides, giving courses, and up-dating the field sales force. Each salesman will spend over 20 percent of a year's time in training alone.

Almost 40 percent of each management center technician's working year is spent traveling to field units from the management center. Each visit is structured to combine on-the-job training for the unit and information gathering. The medical group is involved in extensive testing as well as supervising of training. In addition, the management center carries out certain basic research on diseases such as jungle fever to provide analysis for new drug development. In some cases, such as in Egypt and Kenya, Pfizer research is jointly carried out with the government and national research facilities. This not only improves the efficiency but assures a wider dissemination of Pfizer technology and improved research capacity. Whether managing tests on veterinary or human products, the management center's close involvement with the local scientific and medical community helps provide an important technology link between Nigeria and other African countries as well as creating valuable new information and capability for Pfizer. The management center's basic function is support for pharmaceutical research and production, specifically adapted to Africa and to individual countries.

Value Engineering Applications

At the Nigeria plant, though the plant has yet to come on stream, value engineering assistance has already been received to support existing distribution and feed-mill activities. Special aid was provided to help create the proper sewing machinery and thread for the local feed bags. In addition, a specialized label was developed to help users identify all ingredients. Aid was also received in adapting the local equipment to new energy sources. Special instructions were developed for guides to running agitators, lighting, and all combustion engines.

With the new production plant coming on stream in late 1975, much more on-going linkage will be created to improve techniques, processes and provide support for on-going problem solving and product development.

PRODUCT DEVELOPMENT TECHNOLOGY

The fruits of early investments in technology often are most clearly demonstrated in the more advanced stages of product innovation and development. The stable technology base, once established, soon is stimulated to go beyond improving the immediate process and to strive for new market developments and innovations that might be used throughout the global system. At this stage many plants become exporters of the technology they develop. To accelerate indigenous activities, Pfizer carries out seminars, document exchanges, and technician visits. Regional and global seminars might be held on a standard procedure such as fermentation or purification and pharmaceutical manufacturing. Special meetings are also held on topics such as material management, data processing, or specific production problems. These seminars provide a group atmosphere for exchange of ideas and potential new processes. One such group meeting stimulated detailed research and coordinated product development on specialized tablet coatings, with the results being shared by the entire system.

A special liquid vitamin for animal feeds was developed for South Africa through the collaboration of the Groton Laboratories and New York headquarters. This product will be integrated into other African operations through circulation of manufacturing instructions for the product. Service bulletins, supply manuals, memoranda, and regular manufacturing instructions back up the active telex links that are used daily to quiz different units around the globe for new ideas and solutions for challenging new problems. All of these come together for improving the base for new product development.

In all the different plant environments, the local conditions for medication and legal variations dictate the need for extensive product modification. With a slight change in the quality of local sugar, the entire manufacturing procedure, test specifications, and production guides must be rewritten. This means that even the smallest local dosage plants must have the capability of carrying out their own testing and adaptation since they will have to implement necessary changes. In plants like that in Indonesia, with a staff of over 380 persons (no expatriots), the company had to transfer this basic adaptation technology capability or fail to produce products suitable to the area. Much of the value of the technology transferred can be lost if the full-scale development of quality control, laboratories, and other basic staff are not fully supported. Pfizer depends on the long-run flexibility of its plants and, therefore, accepts this type of technological investment as routine.

Whether designing new packaging for Nigeria or developing an injection form of a medicine normally provided in powder form, the plant will require various outside information and support to fully adapt the new plant to the changing needs of Nigeria. From its initial product advances, new drugs will be suggested and the Nigerian plant will expand its initial laboratory and testing base to include the launching and testing of products specifically designed for the Nigerian market.

The "East Coast Fever" Project in Nairobi is an example of a drug developed uniquely for Africa, tested and cleared through the use of African physicians and managers, and ultimately modified into pill, injection, or liquid dosage by local Pfizer plants to suit the various markets. Nigerian technicians will support that activity and possibly stimulate new drug developments specifically oriented at West Africa or the Niger River region.

ENVIRONMENTAL IMPACT OF TECHNOLOGY

The original distribution activities and later plant development by Pfizer has clearly established production and laboratory technologies in Nigeria. The on-going move toward new products and research promises even more transfer and the development of local technologies. Often the impact of the technology goes beyond the immediate techniques, machines, manuals, and technicians. These impacts depend on the existing technological capability in the supplier and the flexibility of the company in developing new transfer mechanisms. Impact on local Nigerian institutions, standards, and even indirect industries, such as chicken breeding, have been extensive through the association with the emerging Pfizer operations.

Local Suppliers

From the initial contact with the suppliers for construction and engineering equipment, Pfizer must actively work with local resources to improve their quality and provide suppliers with techniques, machines, and financing to stimulate the quality and dependability of supply required in complex manufacturing. Pfizer provided experts to develop the appropriate printing and label production; packaging suppliers also received extensive support to develop new mechanisms and improve quality.

Initial involvement in plastics stimulated contact with several Nigerian bottle manufacturers. In the case of the Puerto Rican plant, where the bottles were often contaminated with bugs and dirt, Pfizer suggested many forms of vacuuming and cleaning until a cost effective solution was found. Pfizer thereby avoided producing its own bottles and containers. For Nigeria, color of glass and plastics were modified in addition to developing sterilizing techniques for the suppliers. Headquarters helped locate Japanese molds that improved container production in Nigeria. In all cases, Pfizer specifications and procedures (quality control and production) are shared with the suppliers to improve their output. Naturally, this means other related chemical and food processing industries will also benefit from the improved supply sources.

Pfizer attempted to develop suppliers of metal packaging in Nigeria to complement those in glass, paper, and plastic. Unfortunately, the lack of metal and skilled tool makers to develop metal shaping devices forced up production costs. Success was gained in developing local carton production, printing, and high-grade sugar supplies. In Puerto Rico, the problem of moisture in the sugar and corn starch required extensive work with suppliers in order to ultimately replace imports. Very high standards are required in the raw materials to avoid sacrificing production quality and efficiency. In some countries, Pfizer is forced to import costly goods that seem to be simple ingredients but cannot be locally developed to specifications.

Introduction of new systems like "shrink wrapping" help suppliers provide a hygenic system for materials storage. This type of plastic wrapping using heat shrinking to form the plastic around the supplies was introduced in Pakistan by Pfizer. Other specifications and equipment were provided to suppliers to help them develop a "pilfer-proof" cap and seal to prevent opening on the shelf and to prohibit moisture contamination. This new sealing technology could be applied to many other forms of packaging. To obtain the right seal, Pfizer quality-control people test the seal quality to assist the manufacturing personnel in the adjustment and operation of the equipment in its initial operation. Special machines may be required to also aid suppliers produce metal tubes for ointments. As in Pakistan and Egypt, Pfizer expects that engineers will have to modify equipment and procedures with the suppliers to reduce the possibility of metal slivers in the tube nozzle.

In Nigeria the veterinary division has already helped suppliers and dealers develop a new feeding block to help Nigerian farmers get medication to range cattle. Raw material suppliers will also be aided in developing quality control for ensuring proper moisture, texture, and other specifications for corn starch, sugar, bone meal, maize, peanut cake, and other basic materials. Pfizer will encourage long-term crop and process planning to help production control and add stability to the local commodity markets. For an oyster shell supplier, Pfizer located and financed the importation of a specialized shell crusher to improve his output.

Agricultural Operations

In Lagos, Ikeja, and three other sites, Pfizer has established small technology centers to support the local animal health activities and feed mills. For animal health, the company has a 26-man staff in Lagos and a field staff of 16 marketing specialists. Veterinarian support is provided by an experienced specialist with three years service in government field research and another three years with Pfizer, including two months veterinarian research and exposure at Nairobi and other Pfizer plants. All animal health personnel work directly with local veterinarian suppliers and practitioners to explain the use and effectiveness of Pfizer treatments. As with the pharmaceutical detail men, these field representatives provide an important link between the practicing veterinarian and new developments in drugs and treatment.

To help develop the market for feed supplements, Pfizer has constructed four regional mills and developed a franchizing system for six additional mills. In the future Pfizer hopes to use many franchisers whom they will support with technical aid and financing. By supporting local business with machinery, loans, and management skills, Pfizer can ultimately concentrate on feed supplements and reduce its commitment to daily management. To ensure proper development of the mills, Pfizer trains the production foreman for three months and the plant engineers for two months and helps prepare the field salesmen for the various mills.

Besides training the maintenance, production, and sales forces of the locally owned mills, Pfizer also has developed unique plant designs to simplify the production process and reduce possible maintenance problems. Since waterproofing in Nigeria of certain plant areas proved impossible, plant designs eliminated elevator pits and other underground construction. Government agricultural specialists work closely with the Pfizer sales representatives and the franchised mill salesmen to ensure widespread distribution, proper use of feeds, and introduction of new agricultural aids and techniques. The collaboration of the Pfizer technology with the local entrepreneurs and government representatives means an increase in feed supplement sales for Pfizer as well as an overall improvement in the care and feeding of livestock.

From its own mills, Pfizer has already lost more than 25 foremen who leave to set up their own operations or to work with emerging competitors. The rapid increase in demand for quality feeds has brought many new mills into operation. At the mill level, Pfizer welcomes the new competitors as it means new markets for its own and other feed supplements.

The local competitors actively seek the Pfizer manuals and procedures for their own plants. Pfizer-required showers and personal hygiene training is spreading to other companies and into homes. Pfizer demonstrated how simple showers could be installed and used in the home and the plant.

A nationwide development in poultry raising and processing

resulted from Pfizer educational and demonstration programs. Since a reliable poultry business depends on reliable feeds, the creation of regional feed mills was the first step for a national poultry industry. After the first plants introduced poultry feed and information on growing, the laying-hen market increased from less than 100 thousand birds to over 2.5 million. Pfizer established demonstration centers to explain how to provide water throughout the day rather than only in morning and evening. Salesmen were prepared in techniques to show farmers how to construct water traps to kill ants that infected chicken houses. Information was supplied to construct cement drains and for improved designs for properly ventilated chicken coups. (Sometimes, improper housing leads to suffocation instead of disease.) Pfizer also printed sample management aids, such as egg-laying record cards, chick-rearing records, and other guides for feed utilization and bird death analysis. Such record-keeping would provide the farmer vital information for selecting proper stock, feeds, and health aids, and ultimately, a productive chicken breeder and grow-out farmer would mean more demand for Pfizer drugs and feeds.

The government worked closely with Pfizer to set up demonstration sites for colony housing and displays on lighting, ventilation, and feeding routines. Nigerians had to be educated on how to gradually switch from growth feeds to laying mixtures. At present, a national poultry association has emerged to continue provision of information. Pfizer aids this group with new guides on complex diseases and other new poultry-raising technology. Long-range planning is also aided by Pfizer forecasts of developing epidemics or disease frequency.

The change in the protein content of the diet of the country resulting from the dynamic growth of poultry raising is best demonstrated in Pfizer's own staff. In 1965, the staff only purchased 30 to 40 eggs a week from the company supply; now the Lagos plant has to supply more than 100 dozen per week. The impact of the poultry-raising technology has clearly benefited Pfizer in terms of feed sales, but it is difficult to measure the impact of this educational technology in terms of local poultry consumption and new farmer income and processing activities.

Institutional and Government Support

The field agricultural programs are but one example of the close association of Pfizer with different government agencies. All of the Pfizer plants and mills are open to the national research institutes and universities. Students in pharmacology, chemistry, and agricultural sciences use the Pfizer facilities extensively. This spreads the technology even further to future managers, consumers, and government technicians.

Government testing agencies are actively involved in cross-pollination of techniques and equipment with Pfizer. As new drugs will be submitted for government testing, Pfizer is encouraged to provide suggestions and guides for procedures and specifications for competitors as well as their own products. In Pakistan, Pfizer supplied a special solvent to the government so that it could isolate the active ingredient in chlorophenical antibiotic syrup.

In addition to local governments, Pfizer actively collaborates with international research sponsored by associations and institutions such as the FAO. In the East Coast Fever Project, Pfizer is supplying technicians and techniques to develop a new application of Terramycin. The development of a long-lasting injection will help prevent certain tick-borne diseases common to Africa, South America, and parts of Asia.

To meet local regulations, Pfizer is developing a network of physicians and laboratories to support clinical tests and evaluations. Many of these doctors had not previously been involved in this type of field research, and the Pfizer association provides information on testing and analyzing new drugs and treatments. The program not only ensures better local drug development but also lends support to the improvement of the national health service network. A local physician in each region is appointed as a medical director to coordinate research at local clinics and labs. The country medical director for Pfizer spends a great deal of his time and that of his staff supporting these local systems to improve the capability of the group to bring new drugs safely to market. These skills will also eventually result in independent basic research and development by the clinics and laboratories.

Along with the support of local universities and standards associations, Pfizer helped found the Nigerian Institute of Management. This local training institution is supported by contributions by Pfizer and by its commitment of over 50 individuals yearly to teach various courses. In addition, Pfizer provides materials and lectures for this center. Pfizer Nigeria also is an active supporter of the Nigerian Veterinary Association and the Association of Agricultural Technology. Individual grants are provided as well as company support of staff memberships. Pfizer periodically also pays for local lectures for these groups and finances certain professional gatherings.

In terms of production, Pfizer makes available films and papers on quality control and production planning. Through the local pharmaceutical manufacturers association and other societies, Pfizer helps build a national attitude toward health care aids and quality production. Headquarters policy counts on long-term education to favor Pfizer's image as a producer of fine products but also as a firm committed to national development.

Philanthropically, the local Pfizer plant gives scholarships annually to three medical students, awards prizes in agriculture and management at five universities, and provides grants for several students in agricultural research. During vacation periods, Pfizer employs an average of eight students a year for training in their plants.

Pfizer's commitment to Nigeria ranges from teaching reading and writing in native tongues to the development of associations and research networks. These forms of technology impact serve Pfizer in the long run by helping provide the infrastructure for better testing and production as well as creating an image of quality and national commitment.

Chapter 9

Pfizer–Brazil

INTEGRATED DOSAGE AND BULK PRODUCTION

Similar to the general evolutionary pattern of international production facilities, Pfizer Brazil, over 25 years, developed to its present state of integrated production and product development activities. The Brazil bulk and dosage plant, with its eight regional facilities, represents a major international extension of corporate production and research activities. The technology base formed during initial sales and later dosage manufacturing served as a base for advanced bulk fermentation and product development. As more technology has been transferred, productivity has increased and the work force has developed accordingly. New products were developed as a result of Brazilian contributions and the local plant is currently exporting bulk materials and finished goods to other Pfizer plants and markets. Evolution of these operations was linked primarily to the development of local markets and an infrastructure capable of supporting advanced pharmaceutical activities.

Historical Development
The first sales of Pfizer products in Brazil began in 1950, and a full-scale marketing and distribution company was established in 1952. In 1953, dosage manufacturing was initiated in Sao Paulo, and agricultural products were introduced in 1957. A rapidly developing health care market encouraged Pfizer to expand its activities

241

and develop a new facility combining both dosage and bulk manufacturing. The plant was relocated to the nearby community of
Guarulohos, and by 1960 the first product completely produced
locally (Terramycin) was marketed. During this same period production of chemicals and vaccines was added. Output of the
Guarulohos plant had doubled by 1967 and a new line of pesticide
products was introduced. Sales figures associated with todays activities and the periods following the first dosage production (1953)
and the integrated plant expansion (1960-1966) are shown in
Table 9-1.

Table 9-1.   Pfizer-Brazil Net Sales, 1954, 1964, 1974 (U.S. $000)

Product	1954	1964	1974
Pharmaceutical	2,667	5,038	17,100
Chemical	—	127	3,600
Agricultural	—	2,332	13,900
Exports	—	—	1,000

Several factors interacted with the growing market to encourage
Pfizer's expansion in Brazil. Corporate experience had been favorable
in overseas production, and the company could capitalize on an
emerging local base of experience and technological expertise in its
expansion. At the same time, the government was encouraging local
production through tax incentives developed in the 1960s for chemical production and later extended to pharmaceuticals in 1970.
Ad valorem taxes were also imposed on imported supplies and intermediate products. The training and manpower development activities in the early Pfizer operation provided a skilled and expanding
management and technical capability within the local company.

During the early stages of production in 1953, the first plant
in Sao Paulo processed injectibles, syrups, capsules, tablets, ointments, and powders. The basic dosage plant employed 320, and had
a field sales force of 80 men. As in the Nigerian case, extensive
training went into the development of the early technical sales group,
and by 1955 the field force exceeded 160. Marketing experts and
training specialists were sent to Brazil from the New York headquarters to prepare the detail men through formal and on the job
training. Even with fierce competition among companies for qualified personnel and a turnover rate as high as 22 percent per year,
Pfizer was able to establish not only a full range dosage production
activity but an extensive technology sales group by 1955.

In 1956, production was expanded to include sterile liquids.

This development, supported by the emerging capabilities of the laboratory and manufacturing specialists, paved the way for the addition of the fermentation, and recovery (purification) technologies required for bulk production of antibiotics. Twenty-two Pfizer technicians, chemists, engineers, and scientists were transferred to Brazil during the period of plant expansion and relocation (1958-1964). By 1965, the Guarulohos management was utilizing foreign nationals only for temporary training requirements. Field units for distribution, testing, sales coordination, and training have spread throughout the country to Fortalieza, Recife, Salvador, Belo Horizonte, Curitiba, Porto Alegre, and Rio de Janeiro. The main plant at Guarulohos presently has 1,300 employees with 560 additional personnel at the field units.

The full-scale activities in Brazil demonstrate the total range of technology that can be effectively transferred. Brazilian product development labs and the medical division are major elements of the global research efforts of Pfizer. Though the economies of scale and availability of scientific infrastructure still demand the majority of Pfizer research to be centralized in Groton, Connecticut, and Sandwich, England, the Guarulohos facility has an extensive range of product, quality-control, and developmental laboratories.

Because of the vulnerability of pharmaceutical companies to losses of years of research, many international chemical and pharmaceutical companies do not transfer their most sensitive industrial secrets. Even in Brazil, with fully developed laboratories and integrated production, few individuals have access to the spores, organisms, or sensitive research that comprise the competitive advantage of Pfizer. Technology for special production techniques and formulations are fully available though they also are somewhat sensitive. Pfizer accepts the risks involved in transferring the production and development technology in order to encourage independent and flexible operations within its system.

Technology Transfer

In the first production developments for Pfizer in Brazil, planning, construction, and design support were provided similar to the patterns of the Nigerian experience. Preparing the field sales forces represented an early obstacle to a rapid plant expansion. The impact of value engineering and product technology is evident in the later stages of Pfizer's activities in the country. Supplier capability and related local infrastructure developed as the Pfizer activities became more complex and extensive.

DEVELOPING THE TECHNOLOGY BASE

In the plant development stages for both dosage and the integrated facility, Pfizer relied heavily on exchanges of personnel to transfer the technology base. All department heads from Brazil received training outside the country in the United States or other Pfizer locations. For the initial dosage plant, a Pfizer manager was sent to establish and run the plant, and carry out training of local personnel. A two-man team in New York was available for periodic visits to inspect and direct specialized activities. Seven different training programs of one to two weeks duration were provided from New York with this team during 1953 and 1954. A foreign expert was also in charge of quality control for three years until 1956.

In the initial dosage plant, air conditioning and plant maintenance proved to be difficult problems that were dealt with by teams sent to Brazil to train mechanics and foremen. During the start-up phase, the general work force available in Sao Paulo was trained not only for installation but also for repair and modification. The delicate processes for sterile production as well as normal production required extensive headquarters support. After the initial construction and start-up problems were mastered, a yearly schedule of visits and exchanges were arranged for the different managers to insure on-going training and introduction of new production techniques. Additional programs had to be developed with the introduction of pesticide production and the manufacturing of agricultural products and vaccines.

Encouragement from the Brazilian government and the newly developed local technical capability spurred Pfizer to begin the planning and design of the new integrated facility as early as 1957. Between 1958 and 1965, headquarters specialists were placed in Brazil to support architectural development, provide civil construction specifications and design the schedules, layouts, drawings, and so forth for the new plant. Building upon the production and laboratory skills developed in the dosage manufacturing stage, the Pfizer staff assembled the materials and personnel for the more complex bulk manufacturing activities.

The new plant has more than a dozen different buildings spread within a 330,000 square meter industrial area. The inital floor space of the production facilities, and administration buildings were in excess of 35,000 square meters. Besides the central administration building, independent buildings house the pharmaceutical plant (dosage operations), the quality-control and spore laboratories, the fermentation and recovery center, the feed-blending center and an extensive utilities center to produce steam, electric power, and brine. A separate facility is also required to prepare and store the specialized mediums in which the antibiotic agents would be fermented. Veterinary and vaccine activities are housed in independent specialized quarters.

By 1960, Brazil had well-developed construction and design capabilities. To augment the basic dosage technology, catalogues, layouts, and specifications were sent into Brazil. The new plant required 830 main drawings, each with 10 to 15 supporting drawings, which involved between 8,500 and 15,000 initial engineering drawings and layouts. Such drawings themselves could not provide the new base technology; specialists transferred to Brazil for developing the new facility had to translate the ideas and concepts represented in the drawings and ensure that implementation was accurate. An office of three draftsmen and several support staff up-dated and carried out engineering changes on a continuing basis. For example, standard production in the integrated plant requires a variety of 44 different test set-ups for each production run that utilize as few as 12 units or as many as 35.

Beginning in 1958 and remaining until 1961 and 1963, two New York headquarters' experts were involved through all production and engineering development. One senior production specialist focused on preparing the overall manufacturing plan and capability while the second engineer concentrated on construction, maintenance, and layout.

After three years with the Pfizer specialists, the Brazilian counterparts assumed overall production responsibility. In the civil engineering and plant engineering area, the headquarters expert was replaced by a Brazilian in 1963, after five years of OJT and direct support of the design and construction efforts.

Due to the complexity of fermentation and recovery in batch production, two corporate staff members were supplied in this area for four years. The expert in recovery also acted as the plant manager from 1964 to 1968 during important periods of plant adjustment and expansion.

To reinforce the sense of job confidence already developed in the expansion of the dosage operations, Pfizer concentrated mainly on training that minimized formal classroom activities. On-the-job training and parallel job responsibilities proved most effective, with extensive documentation support from New York. Teams of three and four engineers and foremen were sent to help install, test, or supervise the first operating runs. These visits normally lasted three to four weeks. To complement the local training, key managers received extensive OJT at other Pfizer plants. The industrial director spent over seven months of extensive training at Pfizer's largest dosage plant in Brooklyn in preparation for handling the specific production problems of the new plant.

At the level of director, manager, and department head, there is an emphasis on recruiting experienced individuals and then complementing them with on the job training. Sixty days of on-the-job training may be required over the six-month introductory period. General workers, supervisors, and chiefs of production lines also receive significant on-the-job training as they increase their responsibilities. The most actively trained and up-dated member within the operation is the detail man (pharmaceutical salesman). Since these individuals are the major link with the outside consuming community, their familiarity with the product characteristics and the related diseases must be extensive. Overall, in a stabilized facility with low turnover such as Pfizer–Brazil, the overall experience requirement, initial training, and OJT would be as shown in table 9–2.

The plant currently experiences a limited turnover of 1 to 10 percent to the outside, which requires an average of 4,000 mandays of training per year for replacements. However, the advancement opportunities within Pfizer Brazil also demands extensive training to replace individuals moving up within the organization. Including turnover to the outside and upward, the company must deal with an effective turnover rate of 5 to 15 percent and supply 8,000 to 10,000 man-days of training to continue its activities. (Example calculation shows an estimate of 8,648 man-days linked with combined turnover figures; see Table 9–2.) In the past the levels of turnover have averaged much higher, with a 20 percent figure to the outside more typical of the early production years. At one time, almost 85 percent of the laboratory personnel were successfully recruited away by an emerging competitor.

Though the managers and senior professionals receive the most extensive outside support in terms of training, this know-how must be passed on to the foremen and skilled laborers to effect a complete transfer of technology. The supervisors and foremen become the internal link between the workers and the professionals to ensure training and manpower development. A completely integrated pharmaceutical plant depends on three types of skilled laborers: industrial workers, chemical laboratory workers, and biological laboratory workers. These direct laborers support the agronomists, pharmacists, veterinarians, and physicians who make up the staff of the production facility.

In the production area, the general worker must develop extensive

Table 9-2. PFIZER-BRAZIL Indirect and Direct Labor, Training, Turnover, OJT, and Initial Experience Required, 1974

POSITION	NUMBERS	ANNUAL TURNOVER %	EXPERIENCE REQUIREMENT		PFIZER[a] TRAINING			Average Days on Fgn Trips or Visits	Estimated Training for Turnover
			ED	Formal Exp.	Formal	OJT	On-going		
Directors	6	10	MA	15	10	60	30	15	69
Managers	82	13	BA	5-10	10	60	10	18	1045
Department Heads	31	13	BA	5-10	10	60	0	18	355
Supervisors Plant	31	15	BA	3	10	60	0	0	326
Sales & Admin.	47	15	BA	3	10	60	0	0	494
Chiefs of production lines	18	15	HS	0	10	30	0	0	108
Professionals									
MD	14	5	MD	5	20	60	5	20	74
Chemists	16	10	MA	4	20	60	5	5	144
Pharmacist	11	12	MA	4	20	60	0	0	106
Agronomists	8	12	MA	4	20	60	0	0	77
Veterinarians	12	10	MV	3	20	30	5	5	72
Engineers	24	12	MA	5	20	220	5	50	850
Production	1	12	MA	5	20	0	0	0	3
Maintenance	3	12	BA	4	5	5	0	0	4
Attorney	1	10	UB	5	20	20	0	0	4
Business Admin. & Economists	38	15	MA	2	20	20	0	0	228
Salesmen Pharm.	210	10	BA	0	60	24	40	0	2604
Ag.	75	15	BA	0	60	24	40	0	1395
General Workers	517	9	HS	0	1	8	2	0	512
Others	180	9	HS	0	1	8	2	0	178
Totals	1325								8648

[a]Complete man-days of training.

experience with matching and shaping of plastics, steels, and other materials. Pipes must be fabricated and fitted and all the workers must have good understanding of measuring and control operations. Specific experience is critical for assembling the varying machines that make up the production stages. Hygiene, both personal and plant, is also another emphasis for training.

Laboratory personnel require much more specialized preparation. Since quality and production control are the mainstay of the production facility, there must be complete confidence in the capability and performance of these individuals. Normally, a good chemical laboratory technician will require two to three years of training before he can independently implement experiments, analyses, and tests. The more advanced technicians and apprentice chemists must be familiar with a wide variety of special fields such as analytical chemistry, mineralogy, physics, and thermodynamics.[a] The effective years of apprenticeship to fulfill some of these different technical areas ranges from two to five years, as follows:

Vocation:	Years Apprenticeship:
Biology Laboratory Technician	3-5
Chemistry Laboratory Technician	3-5
Physics Laboratory Technician	3-5
Instrument Mechanic	3
Chemistry Laboratory Worker	2
Skilled Chemical Worker	3

Production and laboratory technology has high visibility, but the efforts of the detail men, or medical salesmen, represent an equally important technology transfer and critical element of the pharmaceutical operations. At Pfizer Brazil, 210 detail men work in the area of pharmaceutical sales and another 75 in agricultural and veterinarian products. These specialists (most are trained pharmacists) are continually in contact with local physicians and drug outlets to communicate aspects of Pfizer products and the diseases they deal with. This sales link in effect provides the health care centers in Brazil with information and developments on a wide variety of treatments and diagnoses. To effectively sell the pharmaceutical products, the company must develop both documentation and presentations that can effectively communicate the qualities and limitations of the drugs. The physician must also be made aware of the different disease characteristics, drug side effects, and methods of treatment.

[a] Alfons Haussler, "Qualifications and Training," Good Manufacturing Procedures (Zurich: International Federation of Pharmaceutical Manufacturers Association, 1972), p. 312.

Once hired, the salesman is put through a full month of classroom training that utilizes techniques developed at Pfizer headquarters. The second and third months are devoted to on-the-job training, with on-going training of at least 40 and normally 50 man days per year. Each salesman must receive annually a series of ten two-day sections of instruction plus five intensive training courses of five days. Each salesman will be given open- and closed-book tests to ensure full transfer of the complex materials. Each year approximately five new drugs or products must be added to each salesman's catalog. Specifications and recommendations for manuals and brochures and sales techniques are all supplied from New York. General materials that must be covered for all drugs include: drug description, pharmacology, methods of assay and testing, quality control, treatment dynamics, mode of action, side effects, adverse reactions, possible dangers, product research data, and disease prognosis and diagnosis.

The training and management of the field force is the responsibility of the medical division of the plant. In addition, this division is responsible for testing and certification of drugs for the Brazilian market. Responsibilities within the operation are allocated according to function. In Brazil, the organization is headed by a country manager and two executive vice presidents who report to him. The various functions are divided as shown in Table 9-3.

Manpower development also includes extensive programming for general management training. In 1974, 38 staff members will receive specialized courses in 42 different topics. Industrial engineering, microbiology, statistics, marketing administration, and financial administration are examples of the local management programs.

Quality Control

Quality and reliability of the ultimate product depends on application of a system integrating initial research, product development, and finally production and distribution. Even in the earliest stage of simple distribution activities, the transfer of materials and individuals related to quality control is extensive. For Brazil, the development of the quality-control laboratories in the early dosage plant and the later expanded activities in the integrated bulk and dosage activities represent basic steps toward developing the scientific and technical base for later product development and research.

Quality control focuses on three stages of the production cycle. Raw materials are controlled by establishing proper specifications prior to utilization. Continuous and multiple sampling and testing ensures that the raw materials are acceptable for processing. Headquarters manuals and procedures provided for developing the local tests, installing the proper equipment, and maintaining the records and samples.

Table 9-3. Pfizer-Brazil Organization and Manpower Distribution

Country Manager

Executive Vice President	Executive Vice President
Production Manager (66)	Public and Industrial Relations (90)
Vaccine Plant	
Veterinary Production	Manufacturing Division Director (466)
	Basic Prod. Manager
Agricultural Division Manager (140)	Pharmaceutical Prod. Mgr.
Pesticides Sales	Quality Control Mgr.
Veterinary Technical Manager	Engineering Mgr.
Marketing	Process Development Mgr.
Vet. Regional Sales North	Administrative Coordination
Vet. Regional Sales South	
	Administrative Director (179)
Business Manager (12)	Controller
	Treasurer
	EDP Manager
	Materials Mgr.
	Systems and Audit Supvr.
	Pharmaceutical Div. Director (264)
	Medical Director
	General Sales Manager
	Marketing Manager
	Chemical Division Mgr. (3)
	House Counsel (4)
	Consumer Business Director (1)

In-process quality control deals with testing and evaluating the formulas and ingredients. It is important that the system can track all components through the entire procedures and maintain the identity of each batch. Checks are constantly run for weight, moisture, viscosity, acidity, and other specifications.

Quality control over finished goods focuses on testing the stability, toxicity, microbiologic characteristics, and other specifications of the product. Potency and pyrogenic characteristics are also tested in the finished goods before release is approved. Administration and control of these tests is an important management function so that investigation of records or recall of any particular batch can be carried out immediately if necessary.

For example, in the production of Terramycin in a form suitable for intramuscular utilization, all major test categories are applied. Assaying and sampling begins with the raw materials to assure proper potency and characteristics. At each blending station, the homogeneity and moisture must be checked. The weights of the mixtures are tested every fifteen minutes throughout the two-day production run. Potency must be checked to include acidity and the transmission characteristics. Sterility must also be determined along with the pyrogenic characteristics that are evaluated by using live animals and monitoring temperature and vital functions. At the finished product stage, the various specifications established in New York must be assured. The physical characteristics, chemical make up, and biological activity must be determined along with the possibility of heavy metal presence. All of these different control tests are rigidly established in the manufacturing instructions and implemented by the independent quality-control manager. His authority to release satisfactory lots, to reject unsuitable lots, or to recommend recall provides an independent responsibility to complement that of the production manager.

The present Pfizer Brazil plant has a quality-control laboratory consisting of 49 staff members. This group provides quality control support for all activities except vaccine finished goods. The group is comprised of ten pharmacists, three chemists, eight technicians, three pharmacy students in training and 25 laboratory workers. New York has provided the facility with a library containing manuals on test procedures, finished goods specifications, and raw material specifications. In addition, continuing up-dating is ensured through catalogued memorandum on manufacturing changes and adaptations. Direct telex connection with New York and other facilities also provides a linkage for immediate questioning or problem solving on a global basis.

To develop proficiencies with the different techniques and equipment, training is extensive within the quality-control unit. Technicians will go through four months of training with a declining rate of on-the-job instruction. Use of chemical analysis, balance weight bacteriological techniques, spectometry, and many other skills must be mastered. Of the $600,000 spent at the Brazil plant on quality control, over $30,000 is used in training programs. All quality control staff spend at least 3 to 5 percent of their time each month in on-going training; companywide training presentations are offered bi-monthly.

At New York headquarters over $10 million is spent to support the management centers and the field operations in quality-control literature, training, and technical assistance. Each year at least four visits are made to the plant, and several Brazilian technicians and the quality-control manager are asked to visit New York and other Pfizer plants. In addition, Pfizer New York publishes regular reports on quality-control procedures and annually sponsors a series of international seminars on quality control.

Each segment of the production activitiy forms a distinct type of specialization and technology foundation for future expansion. The dosage activities and quality-control laboratories helped support the later addition of fermentation and clarification activities. Introduction of manufacturing of pesticides, agricultural, and veterinary skills bring other specialties and procedures. One of the new areas brought into the new facility is vaccine production. Unlike fermentation of antibiotics, the vaccine process may or may not kill the active organism to develop the necessary antibodies. Success with veterinarian products encouraged Pfizer to emphasize vaccine activities in Brazil, even though it had not been successful in its development activities in the United States. For Pfizer, the only successful work in vaccines had previously been in the United Kingdom.

VALUE ENGINEERING TECHNOLOGIES

During the development of the initial dosage plant and the later integrated facilities in Guarulohos, initial implentation of technology through plant design, construction, and start-up was followed by efforts to improve the productivity and cost effectiveness of the operations. Continuing support from the headquarters in terms of new equipment, antibiotic producing organisms, and technical visits were designed to produce improvements in these areas.

Instead of the long-term exchanges used to develop the base technologies, shorter visits and technical team support continue the on-going technology evolution. From the management center in Coral Gables, Florida, Pfizer–Brazil receives approximately 14 visits each year to routinely support the different division specialties. During 1974, production and maintenance support was provided in two one-week visits and a three-month training visit by a chief engineer from the management center. The pharmaceutical division received six one-week visits to deal with on-going questions of testing drugs for local approval and new techniques for testing. Personnel, legal, agricultural, and quality control all received support in terms of one- and two-week visits from the management center.

To complement the support coming from Coral Gables, each of the six directors spend six to eight weeks out of the plant visiting New York headquarters and attending special task forces and seminars. These trips average three per year per executive and include annual seminars on fermentation, engineering, refining, and pharmaceutical production. In 1974, such topics were covered at a seminar held in Nagoya, Japan.

Pfizer-Brazil, as a major production center in Latin America also received visits in 1974 from several plants in the hemisphere. Argentine, Peruvian, Mexican, and Venezuelan technicians were trained at the Brazil plant for periods ranging from one week to one month. Within Brazil, Pfizer spends $1,400,000 on travel and on additional $200,000 for travel outside of the country. Pfizer's management center in Coral Gables has a staff of 42 to support Brazil and its other Latin American units. After scanning by the management center, specialized documents and at least forty magazines are sent to each facility. An up-to-date and complete production and technical library is considered essential for all operations.

Such on-going technology support has resulted in many improvements in productivity for Pfizer-Brazil. New competitive capability has led to growing exports outside of Brazil of bulk antibiotics, and new techniques and products developed in the country are being sent to other units for consideration. The combination of new equipment, improved techniques, and availability from central research of new organisms leads to improvements in product development.

Specifically in the area of Terramycin, the laboratories at Sandwich created a new culture media that induces more effective growth of the antibiotic producing organism. New York, Groton, and Brazil collaborated on developing new temperature cycles, feed cycles, and changes in the air mixture to increase productivity of fermentation. Brazil also perfected a technique in recovery that increased the yield once growth was completed. This Brazilian development was linked with other new recovery methods evaluated at seminars and through exchange of documentation. The result of these activities in fermentation led to a 700 percent improvement in productivity. Without the new technology inputs, Pfizer-Brazil would have had to have 42 tanks to ferment the same amount now produced by eight tanks (see Tables 9-4 and 9-5).

Table 9-4. Pfizer-Brazil Fermentation Productivity, 1961-1975

	1961	1970	1973	1975
Number of Employees (Fermentation)	25	39	34	36
Plant Productivity[a]	25,000	190,200	217,900	281,000
Productivity per employee	1,000	4,879	6,408	7,805

[a]Productivity measured in kilos of finished antibiotic.

Table 9-5.   Pfizer-Brazil Tank Requirements, 1960-1975

	Production	Productivity	# of Tanks	Production per Tank
1960	22,000	3.1	4	5,400
1975[a]	281,000	16.0	8	34.900

[a]At 1960 productivity level, 42 tanks would have been required in 1975 to produce 281,000.

With increased technology support and improved skills, the Pfizer plant was able to gain in productivity without having to expand staff dramatically. Twenty-five workers were required to support a four-tank operation in the first bulk production in 1960 while only 36 were being used to manage eight tanks in 1975. Productivity per employee had grown from 1,000 kilos to 7,805 kilos of finished antibiotic. The cost effectiveness of this not only meant more success for Pfizer within Brazil, but lower price for the consumers and a new export market for the country in Terramycin and tetracycline.

The same type of efficiency can be seen in other parts of the operation. With a 300 percent increase in sales in most products, Pfizer technical personnel in key plant activities could be reduced by 34 percent as a result of productivity improvements (see Table 9-6.)

Table 9-6.   Pfizer-Brazil Plant Staffing, 1963 and 1974

Staff Level	1963	1974
Plant Management	7	5
Fermentations Operations	42 (4 tanks)	36 (8 tanks)
Recovery Operations	89	33
Pharmaceutical and Agricultural Production	357	222
Engineering	120	93
Laboratories	79	81
Chemical and Vaccines	89	41
TOTALS	783	511

Note: Does not include administration.

Warehousing techniques and inventory control has been another area of technological improvement for Pfizer Brazil. The first improvements in the new plant were made by a two-man team sent to Brazil in 1962 for two months. At present, another renovation of the production planning and inventory control activities are under way. From New York headquarters an electronic data processing expert with twenty-one years' experience in specialized systems design is on loan to Brazil and the Management Center in Coral Gables for two years. With over 700 products, 100 cost centers, 7,000 inventory items, and 6 branch offices, the Pfizer–Brazil operation requires an integrated and effective planning and inventory system. The system developed in Brazil will probably be later shared with Mexico and Argentina.

Within the growing electronic data processing center, there are thirty programmers, analysts, and operators. With access to corporate manuals and systems, the average programmer can become fully effective within a year while between one and two more years are required for qualification as an analyst. This type of administration and planning technology represents the most advanced management technology within the pharmaceutical industry. Turnover is high, nearly 20 percent due to the competition from other companies to hire these specialized technicians. Pfizer circulates these people from Brazil to the other plants in Latin America to increase exposure and exchange of other data processing techniques. In 1974, four of the analysts spent one month each in different locations.

Pfizer–Brazil has concentrated on developing management as fast as possible to match the growth of the plant and to provide individual motivation within the firm. Turnover has been reduced because of significant promotion opportunities though there is fierce competition for Pfizer personnel. The agricultural production manager started with Pfizer as a veterinarian in 1965, was promoted to technical assistant in 1966, to technician manager in 1967, and then to head department in 1975. Many managers began as skilled workers and moved into management with Pfizer training and educational support for outside studies. The manager of basic production began with the plant in 1966 as a recovery assistant. He was promoted to recovery department head, production assistant, refining manager, and head of the department in 1974, while simultaneously studying outside the plant for formal accreditation. The chemical division manager underwent a similar growth by beginning as a package assistant in 1961 and moving up through package supervisor, purchasing assistant, chemical sales supervisor, chemical sales manager, and Chemical Division manager in 1974. He also completed his formal training as a chemist at the same time with Pfizer support.

In the area of production, support came not only from the management center but other Latin American units. In one particular test, the Pfizer–Brazil assay was showing contamination. Extensive review of the procedure could show no cause or explanation.

Fortunately, the Chilean plant had encountered a similar problem and had devised a specialized washing technique for the glassware involved. It required twelve separate washings with a special compound to remove any contamination, and the glass used in testing Terramycin could never be used in the assay in question. The ready availability of the telex leads to two or three such exchanges weekly between the plants, headquarters, and the management center.

Other types of on-going improvements come in the form of purchasing aids and packaging improvements. Within Brazil, purchasing has been developed at the Guarulohos plant supported by global searches from New York and Sandwich, England. These systems and a wide variety of packaging improvements have been supported from the management center.

Overall, the more experienced scientists and technicians feel that about 10 percent of the documentation they receive is new knowledge, but for the staff in training, over 50 percent represents new material. For the senior staff, the seminars and exchanges represent the critical means of on-going technology exchange and professional development.

PRODUCT DEVELOPMENT TECHNOLOGIES

The technology that was transferred into Brazil for the two different plant types established a base of skills and expertise that could be applied to the more complex activities of product modification and development. By the mid-1960s, the Brazilian operation with its extensive laboratories, field medical respresentatives, and production capacity was actively supporting and collaborating with the global research and development activities of the corporation. Specialized market characteristics and local health care systems also played a part in encouraging the expansion of scientific activities in Guarulohos. An independent product development laboratory reporting directly to the management center was established in Brazil, and major research programs were allocated to the country.

Though Pfizer International had not developed the strong research base in vaccines that it had in fermentation, the success of the Brazilian agricultural and veterinarian activities encouraged the corporation to begin vaccine development for all of Latin America in Brazil. Veterinary vaccines were economical in Brazil where indigenous research had not evolved unique products for the local livestock population. Pfizer's early commitment to production of vaccines for the country brought about the creation of a small plant to produce vaccines for poultry, rabies, hoof-and-mouth disease, black leg, distemper, hepititus, and leptospirosis. Pfizer head-quarters assigned one of its leading virologists from England to Brazil to help set up a coordinated facility to radically change and improve the production and development of animal vaccines.

The new vaccine program links five different Latin American plants in an organized research and development activity on animal vaccines, supported by the New York headquarters, Coral Gables, and the animal vaccine group at Sandwich, England. To complement the Brazilian capability, the corporation acquired two small facilities in Mexico and Argentina with expertise in different vaccine areas. The combined manpower represented a larger and more diverse organization that could attack the research problems with support of the Pfizer global system.

The visiting manager for the new vaccine program had spent over fifteen years in this specialized field and previously had been awarded a degree in biochemistry and a Ph.D. in virology. Over ten years of his industrial career had been in management of research and development with Pfizer. At Brazil, he focuses on bringing all the elements of the Latin American system together through a specific research planning system, monthly reports, direct letter and telex links, and several annual meetings of the different groups. New York provided a headquarters coordinator who would travel to each site at least twice a year to coordinate and evaluate the programs.

The major problem for the vaccine program was the development of more effective ways of utilizing industrial production techniques for the vaccines. The combined human resources in Brazil and the other plants would concentrate on different production goals, with Brazil and Argentina well prepared to expand their research on Aftosa (hoof and mouth): Venezuela, on poultry vaccines: and Mexico, on specialized formulation problems. The results of these coordinated programs will provide more cost effective and tailored vaccines for the developing worlds of Latin America and Africa. Success will depend on transferring the technology at Sandwich to the new laboratories and the coordinated vaccine project in Latin America. Pfizer is hopeful that, given the growing technical and scientific infrastructure in Brazil and the other countries, they will be able to link this system with the new developments in bio-technology, and synthetic technologies.

In addition to the vaccine product advances, Brazil has also played an increasing role in basic research in certain drugs tailored to trop-ical and Third World populations. An example of this is the evolution of Oxamniquine, the Pfizer drug that treats schistosomiasis a disease found only in the developing worlds of Latin America, Asia, and Africa. Over 200 million individuals are affected by the three different forms of schistosomiasis. In Brazil and Central America, the Mansomi type is found, Haematobium is the type common to Africa, and Japonicum is the variety common in Asia.

The parasite cycle of Schistosomiasis normally begins with the eggs of the parasite being expelled by an infected human. Untreated sewerage allows the excreted egg to be picked up by snails in fresh water lakes. Within the snail the spores continue their incubation and the embryos mature. The parasite emerges from the snail in a free swimming form and then enters a new human host who is in contact with the water by bathing, swimming, or washing. In Figure 9-1, a graphic drawing used by Pfizer Brazil to train the public in the diseases cycle illustrates the different links in the life cycle of the parasite.

Figure 9-1. Pfizer–Brazil Training Drawing Depiction of Schistosomiasis Cycle.

Once infected, the human host is affected in a number of ways. Anemia and blood disorders, as well as liver malfunctioning develop. Though not fatal in most cases, the inability to work, study, or function effectively has an enormous impact on the individual and his community. With an estimated 10 million persons infected, Brazil is suffering an enormous loss in economic and human potential. (Possibly between 4 and 10 million man-years of productivity potential annually.)

In order to attack the disease, the life chain of the parasite must be broken. Bayer and Gulf have developed methods for treating the water and destroying the intermediary host by killing the snail. This process is expensive and requires widespread and continuous usage. Government experiments in Brazil and other areas have been unsuccessful, and sometimes deadly, in terms of treating the infected individual. However, Pfizer with significant Brazilian cooperation, over a twenty-year period, has developed a one time treatment. Oxamniquine, that can successfully cure most individuals of schistosomiasis.

In 1955, a future Pfizer scientist named Ray Forrester was first brought in contact with the problem of schistosomiasis as a student in Africa. His experience brought him to Pfizer in 1965 as part of the tropical research group in Sandwich, England.

By 1969, the first trial drug had been developed in the United Kingdom for initial pharmicology and toxicology tests. Dr. Forrester travelled to Brazil to enlist the aid of the Brazilian plant and the field medical directors of Pfizer in that country. Dr. Forrester collaborated with two Brazilian physicians in Belo Hoizonte who had been working for years on the problem of a proper schistosomicide. With their help and support of the other field medical directors, Pfizer created a protocol for testing and modifying the proposed treatment. The problems of acute toxicology and chronic side effects required over two years of testing to evaluate.

By 1973, the volunteer human tests had been completed and an enlarged sample of over 1,000 infected Brazilians had been tested with a rate of 97 percent cured through one intramuscular injection. An oral dosage was also developed in Brazil that resulted in a 95 percent record of complete cure.

Given the geographic and health system problems of Brazil, the availability of one dose in either oral or injection form of a schistosomicide was a significant achivement. By 1974, the field force was trained in the basic information that would be passed on to physicians and clinics throughout the country. Pfizer estimates it will distribute over a million dollars worth of the drug in 1975 and make a major contribution to the control of the disease.

Joint public education programs will be developed with the government to complement the information support of local physicians and treatment centers. Work is also underway for modifications of the drug suitable for use in Africa and Asia. Pfizer Brazil's developments for production, testing, and packaging of Oxamniquine will be used to provide the base technology for other Third World plants that will be manufacturing this new drug.

The expansion of production and product lines in Brazil led to the creation of an independent product development and evaluation laboratory. This specialized facility, reporting directly to the management center, was established in Brazil to extend independent support to product development and modification activities. The new group began operations in 1969 and is now manned by four pharmacists, three chemists, and three technical auxiliaries. Their sole function is to develop new products suitable to Latin American needs. Studies are carried out on different raw materials, and new formulations. Specifications, procedures, packaging, and new control techniques are developed by this specialized laboratory. Various projects are coordinated through the management center with the other product development groups at the Latin American plants.

Adapting basic ingredients for different markets is an on-going effort of the specialized LAMC laboratory group as well as the plant specialists. Brazilian laboratories developed the first intramuscular use of Terramycin and a balmintine granular mixture suitable for treating range cattle in Brazil. Previously, the cattle had to be rounded up for intramuscular injections. Pfizer technicians developed a means to granualize the product and then coat it so that the cattle would consume it with the salt left on the range. This modification is now being considered by Pfizer for worldwide distribution. Adapting the mineral supplement for cattle is just one example of how the basic ingredients of Pfizer pharmaceuticals must be constantly adapted for different areas. The technology involved in developing, testing, and documenting is similar to the basic industrial research and prepares the personnel involved for basic or applied research in related fields.

Between 1965 and 1975, 28 new products were developed in Brazil along with thousands of procedure and process modifications. The daily modification of manufacturing procedures to complement new information or local requirements is supported by the technology base of the plant. Specifications, procedures, and quality control must also be adapted with the raw material change. Flavors, storage characteristics, and even labelling must be adjusted to local conditions and market demands.

Whether actively participating in basic research or in the daily routine of product and process modification and development, the Pfizer-Brazil system is gradually adding new complex missions and technologies to its activities. Independence is not the only goal but rather a better global system, with extensive local capabilities to adapt and support the centralized development of new ingredients and procedures. The result is a growing infrastructure in science and technology for Brazil, more effective research for Brazil and other Third World countries, and an improved capability for Pfizer to carry out global research and simultaneously adapt and modify products for local needs.

ENVIRONMENTAL IMPACT

Transferred technology leads to a variety of local impacts including new products, an improved technical and scientific infrastructure, and a more skilled labor force. Within the pharmaceutical operations in Brazil, the field force of the technical and medical director's programs provide an important flow of information and technology to health care centers and practitioners. Turnover of skilled personnel adds critical management and technicians to other pharmaceutical activities and related industries. Close coordination with the government leads to exchanges of tests, procedures, and equipment. Production activities require extensive coordination and joint efforts with suppliers. All of these indirect technology impacts on the environment are associated with the basic manufacturing and sales activities.

Quality control guides and technical training are provided to the suppliers and other supporting organizations. Pfizer has shared its tank and processing technology with the suppliers of many of the raw materials and basic nutrients. Cartons, seals, chemicals, sugar, and bottles are examples of the supplies that must be produced to rigid specifications. Pfizer supplies the specifications, means of controlling quality, and methods for improving production. Tests and equipment were provided to blood-meal producers as well as local slaughter houses that produce lard for the plant. Bone-meal and Kaolien producers also received equipment, manuals, and visits from technicians to improve their production and quality. Equipment suppliers must also be aided to improve their own maintenance and plant management. Cost accounting aids are supplied in certain instances to help develop better estimating and planning capabilities.

Pfizer still has to run an extensive tool and machine shop since local support has not been capable of handling all the complexities of plant maintenance.

A simple raw material such as distilled water may demand modifications in the producer's methods. Difficult aspects like pyrogene contamination can only be controlled with adequate tests and procedures. In some cases the inability of suppliers to provide quality have necessitated even the importation of distilled water. All the aspects of sealing, cleaning, personal hygiene, and materials handling must be a routine part of the suppliers activities.

Pfizer's quality-control and production techniques are passed on to the government and general public areas as well as to suppliers. Eight to ten government technicians are trained annually in Guarulohos. Tests and equipment are also shared when appropriate. Over 3,000 general visitors come through the plant to gain familiarization with manufacturing and laboratory techniques. The vaccine plant has also been of special interest to the national veterinary association and virologists. In addition, much of the technical turnover is linked to new government programs that need the highly specialized skills developed at Pfizer.

With the growth of the Brazilian pharmaceutical and chemical industries, many local companies have established their basic management by hiring away Pfizer and foreign manufacturers' personnel. The general managers of three rapidly growing local firms were formerly Pfizer technicians. In Sao Paulo alone in 1973, there were 3,004 new companies of all types and 118 in the small town of Guarulohos. Extensive benefits and rapid advancement potential at Pfizer have kept turnover low, but increased turnover to growing local industries is expected.

Pfizer opened its quality control laboratories to competitors and the public to help improve the overall industry capabilities. This effort, along with supporting the development of government laboratories, is part of Pfizer's program to support industry and align itself with the development goals of the country. Several new firms beginning operations in Brazil also rely on Pfizer for information and guidance during their own planning and investigation periods.

In terms of community education, the Pfizer sales force is one of the largest on-going training and communication efforts throughout the country on pharmaceutical usage. With the government, Pfizer also carries out special public education activities such as the schistosomiasis campaign. Pfizer may be supplying an appropriate product and gaining new markets but is simultaneously improving the national health care system.

Pfizer actively supports local associations like the veterinarian association and national scientific commissions. The director of veterinarian production is encouraged by Pfizer to also spend time as the head of the national veterinary magazine. Pfizer also supports the magazine with advertizing and articles. Financial aid and staff are also supplied for national seminars and programs on agricultural, veterinary, and medical topics. A specialized poultry chiller bath with chemicals for sanitation was passed on to the general poultry industry through these mechanisms.

Overall, the close work with universities on new courses, cooperation with the government, and support of local industries through planned exchanges and unplanned turnover make wide contributions to the technological environment. However, the unique aspect of the pharmaceutical manufacturing impact is the role of communication with the physicians and pharmacists. The 51,000 doctors in Brazil are in constant contact with the different technical representatives and rely heavily on this form of convenient information transfer. The physicians involved in Pfizer research, field trials, and other tests develop new specialties that are linked with the Pfizer global research programs. Each month Pfizer also receives an average of 100 different mail requests for treatment information and other aid for diagnosis and prognosis from local physicians.

Finally, some products developed by Pfizer, like a drug for testicle cancer, are so expensive on a commercial basis that the technology and a limited production is made available free of charge to the government for distribution. Some products are being produced in government plants, which Pfizer also helps through technical support and documentation. For long-term growth the corporation must not only find cooperative relationships with the government and the general public, but continue to be a source of credible information and products to the physician and health care centers.

※ *Part IV*

Motorola Corporation

Chapter 10

Motorola–Korea

Motorola's Semiconductor Products Division (SPD) is the largest producer of semiconductors in the United States and was one of the largest in a worldwide market of over $5 billion sales in 1974. It produces over 73,000 device types, covering more than 80 percent of all semiconductor applications. Its devices are used in nearly every type of electronic equipment, ranging from low-cost transistors for portable radios and phonographs to devices for computers and aerospace equipment. Its customers include a variety of consumer products (23 percent of sales), computers (28 percent), industrial uses (28 percent), and government (21 percent).

Motorola has four major divisions—communications, SPD, government electronic, and automotive. Each is a relatively self-contained unit, with its own technical staff—that is, there is no corporate manufacturing staff overseeing the operations of all divisions.

SPD, which is headquartered at Phoenix, Arizona, has several production locations: Mesa and Phoenix, Arizona, Guadalajara, Mexico, Seoul, Korea; Toulouse, France, East Kilbride, Scotland; Nogales, Mexico, and more recently in Kuala Lumpur, Malaysis. A new MOS (metal-oxide semiconductors) Division has a factory complex in Austin, Texas.

Motorola's subsidiary in Korea, Motorola Korea, Ltd. (MKL), is an assembly operation that receives basic elements from other units in Motorola and assembles semiconductors for further processing by sister affiliates before sale to customers. Motorola-Korea was started in 1967 with 300 employees, had 5,000 in 1974, and expects to expand still more. Beginning with the application of simple assembly technologies, it has gradually employed more sophisticated methods and even developed some techniques and materials on its own that are not used in other of Motorola's affiliates. Because of its growth and expansion, and despite its improvement in technical skills, Motorola-Korea is a continuing recipient of technical assistance and new technologies from the parent company. As a result it contributes significantly to developmental objectives of the Korean government.

HISTORY

In the late 1950s and early 1960s, several military officers involved in the Korean War determined to help build a bulwark against communism in South Korea by encouraging creation of an industrial base—replacing what it had lost with the division of the country. Among these men was General Van Fleet, who visited the Department of Commerce in 1961 to elicit its support in persuading U.S. companies of the investment opportunities in South Korea. His mission was in line with AID's objective of helping build Korean industry and exports.

General Van Fleet found a receptive audience in Motorola, which was concerned with the problems of development in LDCs and wished to cooperate with U.S. governmental policy. The obstacles to investment in South Korea were reduced somewhat by incentives provided by that government, and the final decision factor was the existence of low-wage labor that could be trained to perform precision assembly tasks.

The decision to establish a plant in South Korea was made by the Semiconductor Products Division (SPD) to reduce costs and maintain its competitive position around the world. SPD furnishes piece parts and materials to Motorola-Korea from its U.S. operations or those in Toulouse (France) or elsewhere. The materials are received in bond and the assembled pieces exported without any sale locally. All materials are owned by SPD, MKL never takes title. However, support facilities, supplies, and machinery are the property of Motorola-Korea.

MKL's supplies are already tested, so that MKL does not perform the difficult tasks of growing crystals, making dies or chips, or testing them; nor does it conduct the final testing of its own product, though it does perform in-process tests for quality control and rejects those that do not meet the tests. In effect, Motorola-Korea performs the middle 65 percent of the process of manufacturing semiconductors, thereby receiving product equal about 15 percent in value and returning product that requires another 20 percent of total value to be added. MKL's deliveries, therefore, are to SPD and its other affiliates around the world—Hong-Kong, Toulouse, and Nogales—that do the final processing before sale to customers.

Since MKL lies in the middle of the production process, it has limited control over what products it assembles; the mix of semi-conductors and the scheduling of production runs are Phoenix's responsibility. This is done by SPD's three departments: discrete transistors, plastic-molded integrated circuits (bi-polar I/C–plastic) and, ceramic integrated circuits (bi-polar I/C, DIP–ceramic, and MOS). Each has a segment of the MKL production facility at its disposal, and each determines its own needs from MKL without reference to the other two. MKL has no capacity to shift between different semiconductors. Not only are the machinery and processes different but such a decision would involve trade-offs among the three departments within SPD. There is some small capacity to shift operations at the margin between DIP and MOS operations, but even this would have to be authorized at Phoenix. Also, if there is to be a longer-range shift in the mix of products or the type of product manufactured, this would be decided at Phoenix in line with larger expansion plans. In sum, each department within SPD has set up its own facility within Motorola-Korea; each separate line is run by a MKL manager who is responsible to that department but who is also under top management within MKL.

The production process at MKL begins with the scribing of wafers or the breaking of wafers already scribed. (A wafer is a disc about two and one-half inches in diameter containing a large number of very small dies on each of which has already been printed the circuit pattern desired, these dies must be broken from the wafer in preparation for their bonding into the particular semiconductor to be produced; the dies range in size from 15 to 65 mills square.) The dies are attached to the header or strip, which is different depending on type of semiconductor, and wires are bonded to the die and the wire leads to make the desired circuit connections. The die and wires are then encapsulated in either a metal can (transistors), or in plastic, or ceramic. The wire leads are then tin (or gold) plated.

The complexities of the process occur in the precision of placing the die on the header or frame, in wire bonding so as to make perfect and durable connections, in encapsulation to prevent moisture from ruining the device, and in plating, so as to make a connection between the lead and the wiring of the final product in which the semiconductor is used.

Not only is technical assistance required on a continuous basis to solve problems that arise, but it is required also each time production specifications or products are changed or whenever there is an expansion of capacity. The problems of expansion are complicated further by the meshing of operations in Motorola–Korea with the new facilities being established in Malaysia, which will be dove-tailed to some extent with those in Korea. Toulouse obtains some of its requirements from MKL, which must be fitted into the production schedule through SPD–Phoenix, and a new joint venture in Japan will obtain assembled product from MKL.

MKL is run virtually by Korean engineer-managers, since all manufacturing units are under Koreans and the only American is the general manager, who has a financial background. He obtained a resident consultant engineer from Phoenix for a few years to provide on-the-spot assistance as a "counsellor, trouble-shooter, communicator," especially needed in a period of expansion such as MKL was going through in 1974–75. This expansion of capacity will require a repetition of much of the technical assistance provided in the construction stage.

The Korean managers have almost all been "promoted from within" as the company expanded, and nearly all are in their early thirties and have been given such responsibility because of their education and abilities. The first general manager was transferred to Guadalajara in January 1969; a replacement, who remained for over four years and returned to Phoenix, was later made responsible again for MKL as regional director operating out of Hong Kong. The present general manager had been the financial officer from start-up and later was assistant general manager before becoming general manager in 1973. Each of these individuals had several years of experience in SPD operations in Phoenix or elsewhere and thus brought expertise and technical abilities to be transferred to the Korean personnel on all aspects of production and management.

ORGANIZATION FOR TECHNOLOGY TRANSFER

SPD's director of international operations is responsible for all activities abroad and for liaison between the U.S. product lines and their counterparts overseas. Since the major production, processes, technologies, and industrial engineering are done within the production lines in the United States, they must be relied upon to support the overseas affiliates. SPD's organization is shown in figure 10–1.

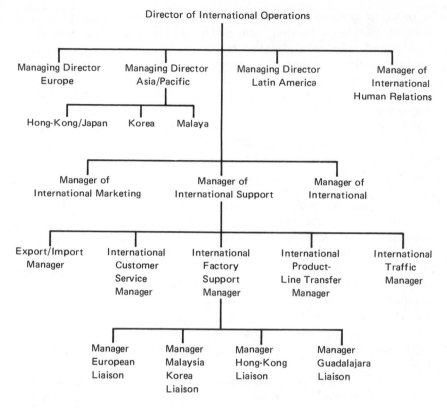

Figure 10-1. Organization of Motorola Special Products Division.

This unit is a major part of SPD, along with the three product divisions: discrete devices, integrated circuits, and MOS. Any new operation overseas would be first conceived by a product line seeking expansion. Discussions would start with the director of international operations as to where a likely location might be or whether existing facilities could be expanded to take on a new line. These discussions would involve the managers of the regional groups and the manager of international support.

The justification for the international support section, since it is liaison and staff, lies in its ability to improve the operations of the product lines, thereby helping a division or an affiliate to meet its goals. When these find that their purposes are well served by the support section, they are willing to see it expanded. The major problem facing the section is that of communication and the willingness on the part of product-line personnel to offer advice to an affiliate. The U.S. personnel are torn between a desire to help and a drive to raise their own profit. Substantial claims on time from abroad can be a burden, conversely, if the problems abroad are not solved, the total division record may suffer. So, international support seeks to marry the interests of SPD with the needs of the affiliate abroad. The responsibilities in the international support section require that each official have from two to five years of experience in production on the line in order to familiarize himself with the problems he will be discussing abroad. Over 90 personnel are in this section effecting liaison with affiliates abroad; more than a third are in customer services, with about 20 in continuing factory support services.

Once a project is conceptualized, it is passed to the controller unit for checkup on budget limitations, to "New Construction" for design, to industrial engineering in the relevant product line for layout, and to the international product-line transfer manager.[a] This last manager formalizes the system and procedure of transfer of the line by tying together the efforts of all departments involved in preparation for transfer of processes and production lines from domestic facilities into overseas factories. He makes occasional visits to the site and may remain for a couple of weeks to provide on-site follow-up to assure that all planned transfer details have been accomplished successfully on a per-product-group, per-country basis.

To facilitate the process transfers, a manual that has been developed on each product line encompasses 23 different topics, including such items as a description of the product line, organization contact list, engineering support, training programs, equipment PERT chart, equipment list, process-flow chart, quality assurance and process specifications, bill of materials, prints, and drawings, layout, shipping and handling, reports, and so forth. The manual for each line consists of one to three looseleaf binders, three inches thick. The transfer manager will hold meetings approximately every two weeks with the eight to twelve officials who are responsible for the various facets of the transfer to make certain that all is going well from the Phoenix

[a]The present transfer manager has been with SPD nearly ten years, coming out of industrial education. He was first a foreman in silicon small signal metal operations and later had general foreman responsibility for assembly and final test; two years later he was assistant production manager in bi-polar integrated circuits plastics with approximately 750 people under him; he was then sent to Korea as a production consultant in the start-up phase and also had advisory responsibility for purchasing and government relations and for training his replacement; two years later he was sent to Guadalajara as manufacturing staff assistant to the general manager and returned to Phoenix in the international factory support group for two and a half years with responsibility for total Asia/Pacific support before assuming his present responsibility.

end, at least. Responsibilities are clearly assigned and target dates set for completion of phases. The transfer manager assumes overall responsibility of all off-shore product-line transfers to better assure minimum production loss, from the time a SPD production department shuts down to ship equipment overseas until they are set up and running in an off-shore factory at full production rates. This also includes covering all technical details involved in support of an off-shore transfer to insure that transfers stay on schedule and are handled in a timely manner.

During this same time of transfer preparation, the transfer manager and the manager of international factory support work very closely together in assuring that the factory support group is totally aware of all "happenings" in preparing themselves for their responsibilities of factory support. These responsibilities start within the transfer preparation period and go into full swing after the off-shore "new" production line is up and running in successful production.

The factory support group is made up almost wholly of engineers, who have come largely from the production or design departments of SPD. Their qualifications must include not only technical know-how but more importantly a desire to help others, for the group does not have its own independent measures of success. These officials are, therefore, extroverts and eager to communicate. The basic technical information comes out of other units, but they must be able to interpret it and even to argue with the originators, for different information and suggestions will come from different segments of SPD through the unit for transfer to an affiliate overseas.

The main burden of the factory–support unit is to facilitate communication, for personnel in the different product lines are changing and have a primary interest in their own (domestic) tasks. The international unit must constantly reintroduce personnel to each other, for they are changed often and stimulate and maintain communication. For this, it receives copies of all Telexes and correspondence between affiliates and the product lines, follows up on action, and pushes for prompt and adequate replies. It, further, performs the function of a diffuser of ideas and solutions by passing them on to other affiliates who might face similar problems.

To be successful, they must be seen by the foreign affiliate as a crucial cog in the wheel and therefore be fully informed of the affiliate's needs. Equally, they must be diplomatic with product-line personnel who do not feel a direct responsibility for overseas activities. To improve their contacts, they visit each plant twice yearly.

An illustration of the unit's work is the plating problem at Motorola–Korea. A Telex came to the bi-polar plastics line in Phoenix, with copies sent to the plating department and to international liaison. Liaison followed up to make sure the response was sent, and a copy of the solution was sent to Malaysia. Then engineers from the line and from plating went to MKL to see whether their solution was applied effectively. Such visits are necessary simply because written instructions on solutions can never be sufficiently complete for another technician to apply fully; face-to-face communications not only provide the final touch but create the trust and familiarity that facilitates future communications. Transfer of know-how, therefore, means not only making specifications available but also making available the men who understand them. There is no single "cookbook" on any product process; the processes are changing so rapidly that Motorola sometimes finds itself building a plant overseas to make products for which not all the technology is fully available as yet. It must, therefore, build in some adaptability while maintaining the highest standards of reliability.

Assistance on reliability and quality assurance comes out of an SPD staff unit that serves all of the product lines and overseas facilities. Motorola seeks to have all products made to the same specifications in all facilities around the world; strict control limits are prescribed by SPD's reliability and quality assurance (R&QA) staff, and each affiliate sends in its control charts for comparison among all plants. The R&QA chief at each affiliate has a dotted-line responsibility to the staff at Phoenix and reports on a series of forms according to policies laid down by SPD. Each affiliate is audited once each year on its procedures, training, environment, inspections, and so forth.

Each affiliate also reports to other affiliates on the results of tests, such as on wire-bonding. Phoenix, however, must be the center for development of many of the testing techniques, for it alone has the necessary equipment. For example, inspection of gauges for temperature, electrical accuracy, and mechanical precision is done at Phoenix, though a recent development is to send out a test lab to each affiliate along with an SPD technician to make on-the-spot tests.

Training in R&QA is especially important, and all affiliates must use training manuals sent out from Phoenix; a training record must be kept on each inspector, and these records are checked by SPD officials on visits. To train high-level personnel, regional meetings of R&QA officials have been held—the only such meetings held within Motorola—covering such topics procedures, standardization, organization, liaison, staff interchanges, budgets, training techniques, measurement techniques, reliability testing, computerized testing systems, failure analysis, calibration of guages, and so forth. Maintenance of proper calibration of guages is probably one of the most difficult of tasks faced by affiliates and often requires technical assistance from Phoenix. But since the test procedures and equipment are constantly being improved, it is doubtful that an operation such as that at MKL could maintain reliability and quality if it were cut off from SPD's R&QA staff.

Another technical service provided by SPD to all affiliates is that stemming from the industrial engineering activities of each major line—discrete units, integrated circuits, and MOS. Each line has its own engineering equipment staff. These groups have several subunits providing direct support to the line operators, developing equipment, constructing equipment, and providing central maintenance. For example, on the discrete line, there are three "direct support" units, each of which has 30 professionals and 70 technicians, for a total of 300 personnel; "equipment development" has 200, of which 75 are engineers, and "equipment construction" has 500 machinists and technicians.

The "direct support" subunits have a tie to similar personnel at Motorola-Korea. The development group designs machines for all affiliates, who receive the same type of equipment on similar lines. And the construction group builds the newly designed machines under contract from the line. The machines are changed in design about once every year, and new technology is built into the machine; for example, the die pick-up, feeding, movement of the head, power, sensing of the strip, and so forth. This group sent an engineer and machinist to MKL to assist in building and installing new molds; trainers have been sent out also.

The engineering support group for the MOS line is not quite so large, that line is relatively new, having grown substantially only in the past two years. It has about 30 personnel in equipment design and the rest of some 300 are in direct support; it uses the construction group in the discrete line to build machines that are needed. Sometimes a new machine is built by an outside company, but it agrees not to disclose the technology built into the machine. Other companies can buy "off-the-shelf" machines, but they are always about three years behind in design and capability. No independent company could buy the same machines as are available to SPD's affiliates abroad. SPD is also ahead in test equipment.

There is no way to measure the value of this equipment design and technology to Motorola-Korea; it would show up in reliability rather than in rejection rates. A company buying "off-the-shelf" machinery would simply not have the same reliability, and the customer will eventually shift away from unreliable devices.

Backstopping these new developments is a "think tank" that is completely separate from machine design but is engaged in "pure research" that will be useful to the design engineers. In addition, SPD relies on the Mechanical Engineering Design Department at Stanford University for assistance on special engineering projects, which completes a progression from "trouble-shooting" of specific problems to long-range development of new processes, products, and technologies.

CONSTRUCTION AND START-UP

The transfer of technology from Phoenix began with the construction of the plant by the Korean Construction Company, whose manager later became an assistant general manager of Motorola-Korea. Even site selection, of course, involves a type of assistance, since officials at Phoenix were responsible for the first decisions to invest and where to put the plant. The plant itself and layout were determined in Phoenix and based on its experience in building similar plants elsewhere. Plant construction and training of personnel were begun in mid-1967. A plant construction supervisor was sent from Phoenix along with an assistant and a person to be in charge of facilities such as water, power, plumbing, and wiring and to train Koreans in their maintenance.

In August 1967, a small building was leased for training purposes, and a large number of Phoenix officials came to train and provide continuous liaison during the year. By the end of 1967, Motorola-Korea had a staff of 300 covering all aspects of the intended operations, these 300 were trained over a four-month period by a production engineer, a process engineer, a maintenance technician, and two production trainers from Phoenix. Special equipment had been brought over for training sessions.

For the installation of equipment in the new plant, a new five-man team was brought over. This equipment was taken directly off of the line in Phoenix to assure the latest production technology for the Korean plant. By January 1968, a flood of personnel arrived from Phoenix to help set up the production lines; some 50 to 60 individuals arrived during the year from all activities within SPD. An engineering consultant was kept in residence from January through August 1968; following him, a production-engineer consultant was provided from September 1968 through April 1970 for the IC area, but he later became consultant to the whole operation and is now handling process production line transfers worldwide. During December 1969 to July 1970, a plating engineer was sent to set up plating shop and to train top personnel.

The construction of this plant brought the first air-conditioned, humidity controlled building to Korea, and the first N_2 plant.

TRANSFER OF PRODUCTION TECHNOLOGY

Considerable technology was transferred not only in the construction phase but also in the start-up of each product line and such transfers have continued on a weekly basis to meet problems that arise. Transfers will increase in frequency and complexity as the facility expands and takes on new products and processes.

Operations

Four distinct operations at Motorola–Korea require different techniques and capabilities: (1) seventeen types of IC-plastic integrated circuits; (2) DIP-ceramic and three categories of MOS (of which there are over 20 different packages); (3) three basic types of transistors; and (4) quality control of all three lines. Workers in each line are trained to do all the different packages, with the best workers shifted to the most difficult packages within the mix being assembled. The mix in MOS is shifting as production in Malaysia expands.

IC-Plastics. The production process starts with scribing of wafers to separate each circuit die from the others. This requires a careful placement of the wafer on a diamond scribing machine, which cuts part-way through the wafer between each die; the wafer is then cracked through by rolling a smooth object over the back of the wafer. There is about a 1.7 percent loss at this stage of the process. Training for this operation requires about two days, and the worker should have at least middle school education.

The dies are examined individually for various metal and other visual imperfections only, since electrical circuit probing has already been done in Phoenix or Toulouse. This process requires one month training and five months of practice to become proficient. About 8 percent rejection of dies occurs at this stage.

Each die is then bonded on a header located on a strip of metal containing also the exterior wire leads; the die is picked up with a vacuum probe needle, placed on the header and eutectically bonded by heat. Training takes one week, with several weeks practice to reduce rejections. Errors occur in the placement of the die on the header, with the die being off center or turned so as to make wire bonding difficult or incorrect. Therefore, each die must be inspected for placement on the header before wiring.

The metal strips, with dies placed on each header, are then passed to the stitch-wire-bonding process, where gold wire is threaded through a capillary and thermo-compression bonded to the die and the external wire leads—the precise connections are made according to SPD's wire bond drawings (bonding diagrams) for the different devices. The specific devices produced are determined by SPD's production schedule for its different customers. The worker is operating a machine which requires 30-power microscopes to observe the work and 15:1 magnification of hand movements to machine movement in the stitching process. Training for this task is carried out according to procedures in a manual supplied by SPD.

The wire-bonding must be inspected to insure reliable connections; training for this process requires three weeks. In addition, wires are sample tested for strength by a destructive wire-pull test. A visual "gate" inspection is made before the batches are passed to the next stage.

The strips are then fitted into a mold for encapsulation of die and wires. Plastic that has been heated is then extruded into the mold to form the plastic body. This process must then be inspected for cracks or voids in the plastic or inadequate sealing. Molding is fairly heavy work, so the girls work in teams of two, spelling each other in the fitting of strips in the molds racks and in the placing of the racks in the extrusion press. The girl at "rest" also inspects the newly molded parts.

After inspection, the strips are electro-plated with tin, followed by a chemical washing, and drying. They are then ready for packing and shipping. The period of processing, including shipment to and from the main Phoenix complex, takes an average of fifteen days for each device.

DIP-ceramic. The Dual-In-Line Package (DIP) integrated circuit differs from the IC-plastic in that it is hermetically sealed in ceramic and the circuits are bonded to a separate set of wire leads, shaped in an inverted U. The die is eutectically bonded to a gold-plated ceramic base, after a similar process of separating it from the wafer by scribing and breaking. Bonding of aluminum wires is accomplished by an ultrasonic weld under similar microscopic observation and ratioed hand movements. Training on the line over about seven weeks is required to achieve proficiency in wire bonding.

The ceramic lid is attached to the base through a heat process—500° furnace held at constant temperature. After this, a microscopic, visual inspection is performed. Further tests such as a temperature helium leak testing, and gross leak testing, using a cycling, bubble test, are performed. Between 0.2 and 0.5 percent rejection occurs at this stage. The wire leads are clipped off and the devices packed for shipment.

The metal-oxide semiconductor (MOS) has a lower power and lower voltage application than the IC or DIP units. Its die is larger (about 240 mills square) and it requires some 40 wires to complete the circuit (compared to less than 6 for most IC and DIP connections).

Transistors. Metal can transistors are assembled on separate lines— a silicon metal line for those that are hermetically sealed and a silicon plastic line. For the former, MKL receives the header, the silicon wafer, the cap (or can), and the aluminum wire. The die is about 1/14th the size of that used in the IC-plastic (i.e., about 15 mills square) and is broken out of a wafer already scribed (the scribing is done in Phoenix by laser or diamond scriber); the wafer is pre-electrical probed, with rejects inked out by Phoenix—sometimes as much as 80 percent of a wafer is not usable.

Given the size of the die, a very precise positioning is necessary on the header, using a 40-power microscope and 15:1 movements of hand to instruments in guiding the process. Only 2 wires are needed to connect the die to the wire leads, done by a stitch-bonding process using thermo-compression.

To make them electrically stable and increase the electrical yield, the pieces are rinsed in pure water for five minutes and then baked to dry out. The cans are finally put on under an electric-weld operation in an enclosed nitrogen-filled glass, dry box to keep out moisture and foreign matter.

The other line assembles silicon-plastic transistors on strips with a plastic bubble to encase the die and wires. After breaking from the wafer, the dies are bonded on the side of a long strip containing as many as 100 transistor headers and wire leads. This is accomplished using a Motorola-designed high-speed machine capable of die-attaching 72 transistors per minute.

The strips are then sample inspected for imperfections in die-bonding and wiring prior to plastic molding. The strips are inspected for leakage and then wire leads are trimmed, preparatory to the final electrical test.

Quality Control. The quality-control responsibility, which is separate from each of the production lines, covers incoming inspection of parts imported, in-line inspection of each product and all processes, and final inspection before shipping. The QA department has two chiefs, who report to the QC manager, and three superintendents—one for each of the three production shifts per day—who are over three supervisors and 92 inspectors per shift. Eleven inspectors in packaging and production control and five clerks bring the total of such positions in MKL to over 320.

Inspection procedures follow precisely those established by Phoenix and are checked against performance criteria also set out by SPD. A looseleaf notebook about one-inch thick provides illustrations and data on all bonds and welds that can be passed or repaired or must be rejected. Rejections are determined against these criteria plus local experience. MKL holds the best record on rejections compared to all other Motorola affiliates in SPD.

Two major problems continue to confront the QA sections over which they have little control: one is the quality of materials inputs, since some slip through the inspection at plants sending them the pieces, the second is the necessity to interpret performance criteria of pieces assembled in the same way as other affiliates of SPD that use MKL product.

When imperfect supplies arrive from Phoenix, Telex's are sent to request advice on correction, replacement, and disposition of the pieces. Therefore, there is rather continuous communication on how to mesh inputs with the production schedules. This is buttressed with a visit from the international quality-assurance officer every eighteen months or so. New information on quality control is also disseminated by memos to all affiliates; if any seem interesting to MKL, it will request fuller reports and will receive a quick briefing by Telex and then the full report for its own perusal.

Additional communication with Phoenix would be required if it were determined that it is necessary to shut down a line because of a rate of rejections that is unacceptable. Though no such case has arisen, it would be up to the general manager to decide whether to ask for someone to be sent out or try to remedy the situation by Telex or phone.

Visits from Phoenix

After the initial training period and start-up, Phoenix has continued to send technicians and engineers to MKL for a variety of problems, at the expense of SPD. Though it is not possible to give a complete record, and it would be too lengthy to print even if a complete record were available, the following steps in MKL growth give an illustration of the type of assistance and a continuing need for it as operations have changed:

Date:	Product:	Visitors:
1967	IC-plastic (14 lead wire type; up to 2000K per week)	10–20
	Transistors (metal can, TO-18, as now produced)	10–20
1968	Transistors (silicon plastic, various dies, TO-92)	
	maintenance technicians	2
	three sets in production trainers	2
	sequence process engineers	2
1968/69	Transistors (silicon metal, TO-5, as now)	1
1970	DIP-ceramic (14 lead, as now):	
	produce engineers	3
	maintenance	2
	process engineers	2
	very high speed die-bonding machine engineer	1
	Tin plating (both IC-plastics and DIP-ceramic)	3
1971	IC-plastic (16 and 24 lead)	
	MOS (1 package; production began at 5 K per week, now up to 20-25 K per week)	2
1972	Silicon plastic line (Uniwat)	1
1973	Silicon plastic ("RF" device)	1
1972/73	Increase in volume of production 30 percent each year	18
1974	Shift of MOS-plastic to Malaysia and soon MOS-ceramic (glass seal)	

In 1973, the EMB bonder was changed to the photocell type, and one engineer came over to install and train operators. In the same year, this engineer, staying one month for the task, also introduced the new soler glass package. In 1974, Phoenix' liaison engineer arrived to discuss the capacity of existing equipment, to review the production processes, and to discuss new developments in molding.

These developments indicate the need for continuing technical assistance, as piece parts are changed and processes improved. These give rise to problems that MKL engineers cannot solve alone; they can solve problems of existing processes, in the main, but would have had to develop their own piece parts to know intimate aspects of the processes involved. Motorola–Korea would also have to have its final production stages in order to be able to resolve all production problems—that is, to be wholly independent of other Motorola units. (Of course, if they were, they would also be cut off from new product and process developments.)

The variety and continuity of assistance from Phoenix can be seen from the following list related to the expansion of facilities, and covering the first six months of 1974:

Technical Field	No of Days
Industrial Engineer (Plant Layout)	28
Production Equipment Engineer	10
Process Engineer (Molding Specialist)	12
Q.A. Engineer	3
Process Engineer	15
Machinist	27
Production Equipment Engineer	9
Production Equipment Engineer	9
Process Engineer	9
Process Engineer	6
Process Engineer	4
Doctor - Chemical Engineering	11
Industrial Engineer (Work Standards)	5
Operation Manager (Product Engineer)	3
Industrial Engineer	2
Industrial Engineer	2
Process Engineer	7
Process Engineer	7
Industrial Engineer (Facility Construction)	20 (3)[b]
Process Engineer	14
Process Engineer (Plating)	11
Industrial Engineer (Facility Construction)	7 (3)
Industrial Engineer (Facility Construction)	7 (4)
Production Equipment Engineer	3
Process Engineer (Molding Specialist)	3 (2)
Industrial Engineer (Facility Construction)	4 (3)
Industrial Engineer (Facility Construction)	11 (4)
Safety Engineer	4
Total =	372

The expertise that is contributed to the MKL operation is reflected in the experience of the present consulting resident engineer

[b]Multiple visits indicated in parentheses.

on the staff. He joined Motorola in 1963 as an electrical technician and went later to wafer processing and then to photo resist, to diffusion, assembly and to the switching-transistor line. In 1969, he went to MKL to assist in the transfer of assembly processes and expanded the TO-18 into different types of packages; it required three weeks to move in the TO-52 packages. At that time he helped resolve a welding problem in the incapsulation process. In a second trip to MKL in 1969, he supervised the start-up of the metal can transistor line during a three-week period.

During early 1970, Motorola–Korea ran into a problem with aluminum wire-bonding; the bonds would break as a result of heat changes in the calculators. Phoenix determined after some testing to switch from thermo-compression bonding to ultrasonic bonding. The necessity to shift was urgent and had to be accomplished in collapsed time. The present resident consultant spent three months preparing for the trip and developing completely new inspection criteria for the line in MKL. In May 1970, he took three ultrasonic machines to MKL and required three weeks to set them up. (Its movements appeared a mystery to the QA engineer in MKL, so the two of them took the machine completely apart to be able to explain how it worked to others.)

In transistors, during 1972, three to four visits were made by Phoenix engineers to check on processes and discuss problems; these were said to be quite useful to MKL in removing various bottlenecks and improving procedures.

In 1973, over 500 man-days of technical assistance were provided to MKL through visits from Phoenix personnel—or one and a half men per day were at the plant on one phase of operations or another. When a new bonding process was put in by SPD, it sent out new equipment and a five-man training team composed of one production engineer, one process engineer, one maintenance engineer, and two trainers. They remained for 16 weeks, or a total of 80 manweeks or 480 man-days (MKL has been on a six-day week).

Also in 1973, an expansion of operations included changes in the molding from phenolic to epoxy compound, which required a new process. Two engineers who came over to help stayed between two and three weeks or for 30 man-days. Since problems continued, many Telexes were used to isolate and solve them; Phoenix eventually sent new circuitry for the temperature control and had to redesign the machine to prevent sticking of the mold. Still later in 1973, two engineers came over and stayed two to three weeks to solve problems of cracking of the mold package on ICs. Over the past two years, seven individuals have been sent to assist in the

IC-plastic production, for about two weeks each. The epoxy process is now as stable and well understood as the old phenolic process.

In IC-plastics, the training specialist who began the initial training in 1967 returned many times in that year to assist. In 1968, several more visits of trainers for a month or more on problem recognition, problem solving, training materials and techniques, and so forth.

In 1971, when the DIP-ceramic line was set up, three Phoenix engineers were sent to install the line, debug it, and train operators. At least once a year since personnel have been sent over for three to four weeks to train maintenance personnel on the line, to change equipment or to improve it, and to train workers to new techniques.

Problems Resolved

Still better illustrations of the assistance provided are the specific problems solved with the help of SPD personnel. One of the production problems in the IC-plastic operation arose when a shift was made from phenolic resin in the molding compound to epoxy, so as to make the package more hermetic against moisture leakage. It was found that "tin whiskers" were popping up when the wire leads were "formed down" at Phoenix. These whiskers were long enough to bridge across and short the adjacent leads. A process was developed for knocking the whiskers off, but it also was not working. Many other tactics were tried, with none succeeding. The problem was believed to be caused by excess resin flash preventing the tin from adhering to the leads. Subsequently, engineering investigations were concentrated on improving the resin cleaning process. The proprietary resin cleaning process was developed by an engineer who was considered the key person to solve the problem; consequently he was dispatched to Korea. After his arrival, a matrix of engineering experiments were run under his direction to isolate the cause of the problem. During the first few days of his visit, an error occurred in the tin plating procedure that produced a plating load of I/Cs in which 100 percent of the parts had whiskers. This being very unusual, created quite a stir. Analysis of these parts revealed that the tin not only was covering the "copper under plate" but the copper under plate seemed to be undercut and was tin plated on both sides. The tin and copper plate were stripped from the leads revealing a venier voil of gold plating. Corrective action was immediately taken by reducing the chemical etching time, which in turn prevented undercutting. Several months later a gold stripping process was added prior to tin plate to remove the gold thereby eliminating the need for strict control of the etching process. This problem could not be solved by the MKL engineers and without a solution, considerable business would have been lost.

After changing to epoxy molding compound, the plastic body began to show cracks. Telexes and samples were sent to Phoenix. Phoenix sent out two engineers for a two-week stay in July. They set up an "eye-ball" inspection procedure. By August 2, a new cleaning procedure had been developed for cleaning plastic residue from the molds, to reduce cracking. It was believed this residue was causing the molded plastic body to stick to the mold. Microscopic inspection was instituted on August 6, as eyeball inspection was not effective in detecting micro cracks. By August 23, they had decided to wax the molds to prevent sticking. October 10, new criteria were established for rejection on the basis of microscopic cracks. By October 16, all molds were waxed more frequently than previously. In November, a procedure was instituted for collecting data on curing times to determine the best duration; somewhat longer periods were found desirable. By December, experiments were set up in MKL to test a variety of causes of the cracks and results were sent to Phoenix. By December, it was confirmed that three different new mold compounds resulted in no cracking. But these new compounds lacking long-term reliability test data had not been approved by all customers. These customers had to be sold on the desirability of a change in compounds—itself a lengthy process.

In December, two engineers arrived in Korea to review the cracking problem as several customers had found cracked product. An "in-process" mold monitor inspection was set up on an hourly basis, to give feedback to the operation so that corrective action could be taken (previously the feedback had been so long delayed that several undesirable batches might be produced). More QA personnel were required, but better information was given operators and more precise mold results were obtained. A decision to slow down the mold opening time resulted in fewer rejections but a slower production rate also; the yield, however, was better. The engineers also found it necessary to limit the scope of operations by the maintenance technicians in order to prevent interference with the responsibilities of production engineers. The production engineers then wrote an "Engineers-Operators' Notice" to instruct maintenance technicians on how to handle cracking problems and then trained them.

In January, through further surveillance, MKL engineers discovered that differences in temperature between the top and the bottom of the mold dies caused cracks, as did misalignments in the top and bottom of the mold dies. As a result of all tests and adjustments, cracking was reduced in January by nearly 90 percent—that is, from 2.0 percent to 0.2 percent of production as a direct consequence of the monitoring procedure instituted by Phoenix engineers, the experiments carried out at MKL, and the continuing technical advice from SPD.

Besides these aids, Phoenix provided three other necessary complements to resolving the problem: it sent four ejector-pin molds (an ejector pin is in each cavity ejecting the mold package from the mold), raising productivity by employing 12-cavity molds rather than 80-cavity molds; it developed a yellow melamine cleaning compound with a residual cleaning agent for use in the 80-cavity molds; and it finally gained approval from customers for use of mold material that did not stick and hence did not crack so easily. None of these could have been accomplished by MKL personnel. The parent-affiliate ties involve, therefore, more than direct technical assistance in order to make technical changes feasible and successful.

Communication on Technical Problems

Besides the physical visits of Phoenix technicians, there is a constant flow of communication by Telex, samples, specifications, new manuals, and test results between Motorola–Korea and counterparts in Phoenix. Communication begins when an operator, utility operator, or production supervisor sees a problem and takes it to a superintendent, or section chief, who cannot resolve it. If the product manager or the production manager also cannot, the process engineer will draft a Telex to his counterpart in Phoenix. Phoenix will then draw on its experience and that at other affiliates to provide the solution, if it can; if not, it will determine the best procedures for finding the solution and will at times hand problems to the Motorola Technical Institute.

Of course, there are a number of problems that Motorola–Korea is now able to handle itself because of its growing expertise. But there are problems of machinery and customer-related services that MKL is not prepared to treat. Whenever it does resolve a problem on its own, it will memo Phoenix for transmission to other affiliates in the event that they run into the same problem.

A check of the daily Telexes showed one or two problems per week were included in all communications, but some quite significant ones were turned up within a span of a few months.

Phoenix will reply to problems posed in two to three days; there appear to be no serious communications bottlenecks, for Phoenix has assigned "sustaining engineers" to respond promptly to requests from MKL.

Despite what appears to be a rapid flow of information and advice, MKL engineers asserted that they could use even more technical information—especially relating to developments in manufacturing technology throughout the industry and what may be happening elsewhere in the company. They feel that regular messages communicating recent developments, even if not directly related to their production lines, would make them more effective than they are presently. They are, therefore, not the least interested in cutting themselves off from any information or technology flows; on the contrary, they would like to increase them. Frequent reports are also made on inspection procedures and results of experiments set up (such as debugging the new high-speed IC die-bonding machine), which will elicit comments from Phoenix.

The three production departments of Motorola–Korea write "weekly narratives" to their counterparts at Phoenix that follow-up on previous communications concerning problems and show any new ones that have developed. Telexes sent on Friday afternoons are received by Phoenix on Friday morning (due to 24-hour lag in time going East), and MKL will generally have a reply by Monday, in time for production start-up.

There are times, especially in new product start-up, when one or more Telexes per day are sent concerning a problem and asking advice, one-fifth of which would suggest an MKL solution for checking in Phoenix. Korean engineers are encouraged to use the "full-time leased" Telex line or, if the problem is important, to telephone. Fast, good communication has been a key element in MKL success. Mail is too slow for a high-volume factory of this type. An example was the fact that the furnace capacity in die-bonding of both DIP ceramic and MOS became a bottleneck with the two production lines using the same facility. The department required instruction from Phoenix as to how to schedule production to meet end-user schedules.

Despite the use of frequent Telexes and the visits that have established a feeling of openness, written communication is often not adequate, and oral or visual demonstrations are necessary. Therefore, phones are used or personnel must travel. For example, MOS counterparts were called on the phone to solve the high reject rate in imported piece parts, for the line either had to be shut down temporarily or continue with very low yields.

Each product group reports problems to Phoenix on various occasions and will have different capacities to handle on their own; in addition, the reaction in Phoenix is apparently different, according to which person receives the Telex for handling. The resident consulting engineer at MKL helps to maintain communication flows, interpret messages, translate between the personalities involved, adopt the modes of expression to the operating modes of individuals at each end of the communication line. Despite these efforts small problems of communication occasionally remain because of the difficulty of each to understand precisely what the other wants; this understanding differs, of course, among managers as well as among departments in each.

As a follow-up on continuing communication, two engineers visit every six months to check on progress and pursue problems anywhere on the three lines. Of particular importance during these visits is a check on inspection and test levels, production performance, and the correlation of production with materials sent from Phoenix. For example, shifts in the supplies of tin plate from the United States or of gold or aluminum wire cause differences in test results; consequently, either the materials have to be changed or a determination made as to whether the new test results are acceptable.

Technical Developments

The engineering background of the managers of Motorola–Korea, the necessity to solve new problems, the assistance by Phoenix, and the need to handle problems without assistance at times has developed within MKL an expertise of its own. This expertise is the direct result of the international technology transfer and the opportunity to employ local talent; it provides another source of technology for transmittal to all affiliate companies.

Some years ago the DIP-ceramic line had a 13 percent rejection rate at the furnace operation; after considerable work by MKL engineers, the rejection rate has been brought down to between 0.1 percent and 1.0 percent on a weekly basis. The DIP-ceramic engineers also designed a carrier for the pieces as they are passed from operator to operator, it is of extruded aluminum and made in Korea, it is now used in Phoenix and other affiliates of Motorola. MKL has also designed, developed, constructed, and tested in operation a continuous-feed lead-clipping machine to cut the borders off of the wire-leads of the DIP-ceramic packages; it is a semi-automatic lead-clipper that requires the operator only to insert the device (rather than insert and retrieve), which speeds up operations considerably. In a two week period an operator at this machine doubled production of former machines: MKL is now planning to add an automatic feed mechanism that will raise production to three times the level of old machines. A process engineer also developed a light diffuser for die-bonding machine that enabled the operators to see oxide colors on the die being bonded; it is now used in Phoenix, Toulouse, and Nogales. MKL has also developed a simple machine to crush molding from rejected pieces so as to be able to reclaim the gold from them. To eliminate a large number of hand movements in loading transistors one at a time from container to tray for capping, MKL engineers developed a simple mechanism for multiple loading. To gain space on the production line, the die-bond and wire-bond machines were compacted to one half their former size by cutting the feed-strip containers in length and reducing the size of the support parts of the machine. The capacity of the line was increased substantially without investment in new space.

MKL engineers can, however, make changes in production procedures and alter the line on its own initiative. They can modify the machines to achieve better performance, as with the circuit bonder and welder guide finger—both modifications were sent to Phoenix for an "OK" prior to a change over. They were OK'd, but MKL does not know whether they were adopted elsewhere.

MKL engineers have, of course, acquired the capability to solve smaller problems such as improper plating thickness on frames, or poor workmanship, or bad-bonding pressures, or to develop new inspection procedures. In addition, MKL was instrumental in solving a problem with the ultrasonic wire bonder. The first efforts directed by Phoenix resulted in new problems; after some modifications by MKL of these instructions, the problem was resolved and the procedures accepted by Phoenix. This modification is now being installed in all such machines in Motorola. MKL engineers also solved a problem of TO-92 molding.

When the company shifted from a seven-day to a six-day week, the nitrogen generating capacity had to be rescaled. This was accomplished by MKL engineers (after one travelled to the Japanese manufacturer to check out his ideas) without major changes in the equipment.

In the IC-P area, it is necessary to clean the molding dies as often as every 10 "shots" (shot is a term referring to a mold cycle). These nonproductive shots utilize a mold-cleaning compound called "melamine" that contains cleaning and release agents to remove residue epoxy left on the mold surface. As mentioned previously, this residue causes sticking problems. Since it is necessary to use a "dummy strip" (a strip with no die or wires) an extra inventory must be maintained just to perform this operation. At times when no dummy strips are available, they are compelled to use gold plated, good strip, which requires additional processing to salvage the gold.

MKL engineering devised a paper insert that could be used in lieu of the strip. This innovation not only represented a tremendous cost savings to Motorola but had other side benefits such as being able to retain the excess resins from the melamine. This innovation resulted in SPD presenting an engineering award to the individual who devised the process.

In some instances Motorola-Korea has had to adjust innovations by Phoenix. (For example, Phoenix developed a mold form that was assembled rather than cast in one piece; MKL had to rescale it to proper tolerances.) There are limits to MKL capabilities, however. They exist in the lack of ability to develop new products or make major adaptations in products or to develop new processes for manufacturing (which capability requires a broader knowledge of semiconductors than MKL engineers have yet acquired); nor could MKL produce the die-wafer or make any contribution to its development or production processes.

As a result of growing expertise in MKL and its contributions to more efficient operations, Phoenix engineers are seen as less adamant as to *how* a particular process should be carried out and are more willing to rely on suggestions from MKL. A closer partnership is being forged out of communication and experience.

MANUFACTURING SUPPORT SERVICES AND FACILITIES

Technology is transferred not only in processes of direct production but also in supporting services, beginning with the construction of the plant, that now extend throughout the production day and that are also available to suppliers both inside and outside Motorola. We have already reviewed the assistance extended during the initial plant-construction phase, but similar assistance is presently being received by MKL as it creates additional capacity.

Technical assistance in the production support services are charged differently from those extended by the three departments of SPD for semiconductor production. If MKL wants assistance in support services from Phoenix, it must pay the expenses of the trip and per diem to the expert sent; it does not pay the annual salary of the individual but, in effect, pays all additional costs incurred by the trip. Someone in this area has been requested each year, on the average over the past five years, for a stay of about three months each time.

Support services for the production lines at MKL include facilities, operation of a machine shop, electrical maintenance and engineering, micrososcope repair time and motion studies, safety and layout changes, and pelletizing of mold compounds.

Machine Shop

Phoenix provided extensive help in setting up the maching shop. One man was sent out for a full year in 1968 to buy tools, run the shop, and get it firmly based. He has returned at least once each year. Now the shop decides its own needs and buys its own tooling. Some eighty machinists work in the shop in shifts, turnover is no more than one man per year (who usually emigrates to the States or is hired by a local company).

The machine shop, fabricated some of the parts for a date-coding machine for TO-92 strips, designed by Phoenix; it has since built three of the machines for Guadalajara and Malaysia. It has also built a die-bonding machine for Malaysia, to Phoenix specifications, and built two sets of very high-speed die bonders for the TO-92 line at MKL.

The machine shop also designed the continuous-feed lead-clipper mentioned earlier. A mold-crushing machine developed by the machine shop was an adaptation of a rack crusher. It is now exported to other affiliates at Hong Kong and Nogales.

The shop makes all spare parts for the machines MKL now uses. For example, the wire-bonding needle (or capillary) must be reworked after each shift, or at least each day. The shop has developed new tools for this and added new techniques of repair to raise productivity from 20 repaired per day per machinist to over 40 per day. It requires eight weeks to train a machinist to repair the needle, which has to have its inside and outside radius retooled precisely. The shop also makes its own forms for holding molds in the heat furnace.

The machine shop has no precise counterpart in Phoenix, so it cannot be backstopped on a daily basis as the production lines have been. For nearly a year now, it has been able to resolve its own problems without help from SPD. The expertise will be used in building parts or entire machines for the Malaysian affiliate, starting with the cutter-blades for wire-bonding and tools for repairing needles. However, the shop still relies on blueprints from Phoenix.

Safety

On safety techniques, there is a counterpart in Phoenix who backstops Motorola–Korea on all new safety regulations and specifications in safety equipment. MKL has been assiduous in following such instructions and has records in each department that are equal or better than those of other affiliates. But constant efforts are necessary to instill a mental attitude on the part of the workers in support of the safety program.

Facilities Expansion

In 1967, a Phoenix engineer was sent with specifications on power consumption, air conditioning, water consumption, space needs, and so forth to set up the support facilities. Various visits have since been made, the most recent one being of three weeks duration in November 1973 to check on facilities and the organization of the expanded support system.

One facility requiring expansion is nitrogen generation capacity, by 25 percent to permit increase of daily consumption that will result from a drop from the seven-day to six-day week. Phoenix recommended either an increase in pressure or a modification of instrumentation, MKL chose the latter.

Another modification being made is to switch from use of city water for cooling equipment to a recycled water supply with their own water cooling tower, thereby saving water and money. MKL must now decide what capacity to make the tower, which requires a coordination of all three production lines and plating both now and in the future; this will require assistance from Phoenix. Phoenix will also have to help on the complex problem of piping of all gases to the production lines and maintenance of these lines. Its engineers laid out the original system, but expansion will alter it and require increased capacity.

The present expansion, which is to be carried out in stages, looks toward a doubling of capacity by 1979. One engineer will spend at least six months each year at Motorola–Korea to work on the central plant, fire protection and sprinkler systems, a chemical storage building, and the expansion which was due to be completed in December 1975. The decisions as to who should be sent are made in Phoenix; the responsibilities of the visitor are not only to extend technical assistance and help make decisions but to see that the plant is expanded within the budget limitations and the specifications agreed upon, he is a liaison between MKL and each SPD department whose line is being expanded there.

During my own visit to MKL, a project manager for international facilities was there checking on progress and making decisions. His contribution is out of many years of experience and knowledge of Motorola's standards, having previously been the chief plant engineer at Phoenix. His relationship with Phoenix is such that he can change the specifications at MKL and SPD will support his decision, so he does not have to check back on every change. A main responsibility is to see that construction stays within the budget; MKL will need such help through 1978, partly because it is understaffed and partly because of inexperience of present staff, but primarily to cut liaison time between MKL and Phoenix. Thus, his principle contribution is "on-the-spot" decision making, plus some technical assistance, also, he can argue more successfully for additional funds, when needed, than could a Korean whose experience and decisions would be discounted at SPD.

But, his technical assistance is also necessary. For example, he was able to determine that a high voltage power station would not need to be rebuilt until 1977, and he gave assistance in the construction of electrical power stations, air conditioning, and water supply. (The kitchen air conditioning had been tied into the office area air conditioning so that kitchen odors wafted into the offices.) He also reduced the size of the emergency light output, thereby cutting costs, and vetoed an MKL request to cut humidity to 35 percent, but "OKed" it down to 50 percent only in the new production area. On one trip a problem of disposal of waste from plating processes had to be solved; they had to heat waste-water in winter to activate chemicals in the tanks to clean the water.

Beyond these specific types of problems, assistance from Phoenix is needed to see that various facilities fit together and that each part interfaces with the rest appropriately.

Suppliers and Affiliates

Not only is technical assistance obtained by MKL from Phoenix, but MKL provides technical assistance in turn to other affiliates and to local suppliers. These suppliers are often the sources of supply to SPD and other Motorola facilities—such as the finger-cotts (rubber protection worn on fingers to prevent contamination of the product in handling) and trays and carriers for the products in process, all developed by MKL (from Phoenix specification) and promoted with local producers.

Local suppliers have been developed for over 200 items required by MKL, included are carriers, trays, boxes, and containers, all of which are made to SPD or MKL specifications. The finger-cotts are now supplied not only to Malaysia, Guadalajara, Toulouse, and Phoenix, but also to Fairchild in Korea, and the Korean producer is now investing in production facilities in Malaysia. Similarly, the containers for transistors were developed in MKL and are now exported by the supplier to Phoenix.

Some 20 percent of the 200 suppliers have been assisted directly in production problems—mostly production control and scheduling. Quality assurance personnel from MKL are sent to the suppliers to assist in identifying problems and resolving them. (Items bought locally for which assistance is not needed include daily services such as kitchen supplies and many construction materials.)

The fact that MKL is a large supplier of the intermediate product for all affiliates means that it receives a number of visitors from other affiliates who want to see what it is doing and how. For some, expecially the Malaysian and Japanese companies, this is a learning experience. In fact, Korea is a better learning resource for Malaysia than SPD in Phoenix in view of the fact that Phoenix is not a large-scale assembly operation and is much farther away.

Knowledge of foreign markets and ways of doing business led to MKL's business manager being asked by the Korean government to go abroad to see whether Korea could export more to U.S. and European purchasers. He also helped the Consumer Products Division of Motorola source supplies in Korea, prior to its sale to Matsushista of Japan.

TRAINING

The transfer of technology occurs largely through training on the job at Motorola-Korea, with only a few days or weeks in preparatory lessons for workers. Higher-level personnel are usually graduate engineers or are encouraged to continue their education. In addition, some of the top-level personnel have visited Phoenix for on-the-spot observation of techniques and processes. Worker turnover means that new employees must be trained on a continuous basis. And promotion policies encourage broadening of experience in the plant's operations to find the best man for supervisory and managerial positions.

Programs

Personnel and training programs at Motorola are under a vice president for human resources, and the line of authority for these activities runs outside of normal managerial channels. The direct (solid-line) authority runs from the vice president to the personnel manager at MKL through the international personnel manager at SPD and a regional manager now located in Seoul, Korea, to whom the MKL personnel manager reports. Each of these officers also have dotted-line responsibility to division, regional, and local managers.

The training programs are the responsibility of MKL's personnel manager. He trains several levels of workers during nonworking hours (usually after the relevant shift). In addition, programs are offered to supervisors, maintenance officers, and utility operators. During 1972, 169 personnel were trained as supervisors. The supervisory course was last offered in August through October 1973, with 204 personnel attending. The distribution of subjects was as follows: motivation (2 hours); supervisor's role (2), persuasion (2); human relations (2), work measurement (2); productivity stimulation (2); and other topics (15) for a total of 27 hours.

Production supervisors development training is given to twenty personnel during the year for a total of 23 hours. MKL instructors are used plus some professors from local universities and training institutes. The content of these training courses is largely developed from experience of personnel managers and instructors within MKL as to how to proceed and what workers need.

Twenty maintenance technicians were trained during early 1974 in courses covering the four departments of the company (including plating). The objectives were to "strengthen mental ability, improve human relations, improve understanding of personnel policies, disseminate Motorola's policies." The program required sixteen hours of classroom teaching and discussion. The subjects covered included: Motorola's contributions to Korea; applications of semiconductors; attitudes towards maintenance, role of maintenance; communications, safety; personnel regulations; and the credit association.

During the first half of 1974, eighteen utility operators (who are the multiple-talent workers and serve as supervisors, can fill in as needed, and who train others) were given ten hours of training in communication and persuasion, the concept of quality control, leadership and human relations, personnel regulations, and training reports. This program is run six times each year for two utility operators from each of the three shifts, for a total of 36 during the entire year.

All workers are given an orientation program to acclimate them to Motorola rules, safety regulations, company organization, and working in air-conditioned and decontaminated conditions. Beyond this, weekly training sessions are available on basic education topics, including public morality (e.g., behavior in the company leased buses that pick up and return all workers daily), promptness, recreation for health and company facilities, personal hygiene, cleanliness in production, thrift, waste, use of rest rooms, and so forth.

Personalities of some "bad actors" have been changed merely by working at MKL; some, who had previously been involved in petty crime, were merely behaving badly, or were drinking too much, have proclaimed that they have gone through a type of "conversion" to understanding that life can and should be lived more cleanly. However, some cannot adjust and simply leave. Others find that they have become more thrifty and can improve their living standards and family life. For example, the provision of free transportation, portal-to-portal, which has cut the aftershift "pub-hopping" that most Koreans do, delivers the men home "on time" and makes wives much happier; such free transportation is not the custom in Korea. Personality development is seen as important for the young girls, and the managerial staff feels a direct responsibility to guide them into better lives and tries to set examples in their own. At times, "refinement" courses, taught by teachers from outside, are conducted for the girls on courtesy, behavior, etiquette, and home-making. And recreation facilities are provided in a swimming pool, tennis courts, avocation clubs, and sports contests, plus vacation sites and facilities.

For those who are stimulated to continue their education, non-college courses and college degree programs are supported by the company. In 1972, 72 noncollege programs were taken by em-ployees, at a total cost to the company of $1,000. These courses covered topics such as English language, quality control, safety, and various technical areas. The company pays 100 percent of the cost of any courses it requests an employee to take and 80 percent of all others. College degree programs were pursued by 64 employees during 1972, at a total cost of $10,000 to the company; of the 64, 34 were in electronics and electrical engineering; 13 were in manage-ment and administration; 15 were in other types of engineering; and 2 miscellaneous. Of the total, 10 were graduate programs, and 54 were undergraduates.

At present, MKL provides no grants to universities for general scholarships, though it is looking into the problems of doing so in Korea; such an action would require Phoenix's approval.

In 1973, the number of technicians and engineers in university courses with MKL support rose to 67, of which 10 remained in graduate courses and 57 were undergraduates. Cost to the company rose to $14,000.

Support of children's education has been offered for middle school and high school, but only four parents took advantage of it for their children. This is explainable partly by the fact that the work force is quite young; almost all in-line workers are single girls and only a very few stay after marriage. However, this still leaves several hundred supervisors and engineers, and given the youth of the company and the fact it started with young staff and that only a few have children in the middle school, much less high school, the figure is expected to rise substantially.

During 1974, an Apprentice Engineering Program was begun, modeled after the Phoenix program. Four young engineers were chosen to be rotated through the areas of "facilities, manufacturing support, the three production departments, and quality assurance." They were in each for eight to ten weeks, which amounted to a year of apprenticeship. Of the four chosen two already had their engin-eering degrees and the other two had nearly completed their pro-grams. During this training period, salaries are paid at the level of a junior engineer.

For higher-level management training, the Motorola Executive Institute (headquartered near Phoenix) conducted a program in Bangkok, to which three MKL managers were sent. In addition, an industrial engineering program was conducted in Malaysia for two weeks in 1974; this was attended by one department manager, one chief industrial engineer, and one line-process engineer.

In addition, the Manufacturing Support Manager and his Engineer went to Kuala Lumpur for a three-week course given by Science Management consultants.

Visits Abroad

In addition to this in-plant program, top-level personnel have visited abroad for training and communication purposes. However, these have been few in comparison to the trips from Phoenix to MKL, in the first four years of operation, there were only 8 "indoctrination" trips from Motorola–Korea to Phoenix. This was a result of company decisions, tight governmental restrictions on travel of Koreans abroad at that time, and the high level of training of Korean engineers. A former assistant manager is now in Phoenix for two to three years. The QA manager had a work and training assignment in quality assurance for three years at Phoenix, during which he concentrated on "reliability quality control," which is beyond the "in-process" control that is largely done within MKL. A supervisor is now in Phoenix with "international quality control." One engineer, who was sent to Malaysia as production manager, worked first in the training lab and later in the plant start-up. These overseas assignments are, of course, also work assignments, but the training will undoubtedly lead to promotions for these individuals in the future since they gain considerably in experience and expertise.

In addition to those mentioned, three other personnel have been sent to Malaysia: one process engineer; one in-process quality-assurance section chief, and one in-line maintenance supervisor. Each will remain for more than a month. The process engineer returned with a desire to learn all three MKL departments because he had not been able to answer all questions asked in Malaysia. Ten Maylaysian personnel have been sent to MKL for over a month to learn Motorola operations.

The I-C engineering chief was in Malaysia three or four weeks last year. In addition, the maintenance supervisor was in Malaysia for six weeks to help set up the TO-92 line (strip-plastic transistor); the Malaysian plant has asked for detailed information on operations.

Among the top managers of MKL during 1974, all departmental managers have been to Phoenix for three or four weeks in order to see the operations there, meet their counterparts, and establish closer relationship. Some have been three times, and some have gone to Chicago to see corporate headquarters operations. In addition, some have stopped in Tokyo to visit the purchasing department there. For example, one line manager made a three-week visit in 1970 to Phoenix to see both discrete and integrated operations; he examined the preproduction processes of getting an operation started as well as the post-processes. As a result, he gained a better understanding of SPD's way of thinking and operation and of the processes and the people with whom he has to communicate. Specifically, he discussed problems of bonding of wire on metal surfaces, washing procedures, and production assembly and scheduling. Prior to this visit, he had not understood precisely what Phoenix wanted; after the visit, he got the instructions straight largely because he could perceive the reasoning behind them. On another trip to Bangkok to the MEI session, he also visited Malaysia, Taiwan, and Hong Kong operations. The last was most important because Hong Kong does the testing of discrete products from MKL; yet he had never met the men there and was not familiar with the way they see things.

The manufacturing manager was in Phoenix during 1974 for two weeks examining wafer processing. Though MKL does not produce wafers, it was most instructive to him to get an idea of the procedures and to see where problems in processing might lie. He felt that the visit should have been earlier in his career and concentrated more on a few facilities, but he simply could not spare the time sooner nor more time than permitted this trip. The chance to see new processes that were being developed (and which MKL will eventually adopt) made him more eager to prepare for them and gave him a greater understanding of the desirability of reducing costs with these processes. He urged that exchanges of personnel should be two-way, with those at Nogales visiting MKL and MKL visiting there to keep in mind the exact processes being used at each.

One product manager would like his process engineer chief and his four engineers to go to Phoenix to meet their counterparts and see the operations; as a result, their greater understanding would mean that he would have less difficulty explaining the need to follow instructions.

The manager of DIP-ceramic examined metal finishing and plating during his trip to the States and observed the plants at Mesa, Mohave, and Phoenix (for MOS). Having seen the operations from the beginning of the wafer, he now could see how his operations fit into the whole and appreciate the necessity of production care and efficiency.

The original group trained to start-up the MKL operation and to manage it were graduate engineers (university degrees), with quite high educational backgrounds. They could be trained, therefore, by sending U.S. engineers to them in Seoul. The program was quite successful, and all of the present top managers (save one) came from this group. Of the original group of 50 who were trained as top-level engineers (managers), supervisors, and engineering technicians, only one-third have left over the seven-year period, and these almost wholly for jobs in the United States or to small electrical companies where they could utilize the skills learned at Motorola.

Contrary to practices in companies such as ITT and Xerox, Motorola has had no "task force" meetings or meetings of specialized personnel around the world save for one session of quality-control managers. The Bangkok seminar was for training rather than for exchanges of experience or presentation of technical papers among colleagues. But it is expected that meetings will soon be held and will draw personnel from the seven major production facilities—Guadalajara, Nogales, Malaysia, East Killbride, Phoenix, Toulouse, and Seoul—on a regional basis.

There is a limit to the efficiency of relying wholly on communications and instructions from Phoenix, however. For example, the MKL engineering staff matured rapidly during the period 1970–71 when the U.S. recession cut personnel trips to MKL and no U.S. resident engineer was provided for a year from early 1973 to early 1974. This period gave them greater confidence in their ability to handle a variety of smaller problems to increase the efficiency of the plant on their own.

Turnover

Training programs are, of course, affected by the rate of turnover and the necessity to train new entrants. If such entrants are likely to pursue employment elsewhere when they leave MKL, they provide an insemination of technology into the economy. In fact, MKL's turnover is the lowest among Motorola affiliates. Working conditions are better than in most any other company in Korea (air conditioning, cleanliness, free lunch, transportation, and low noise level) so that workers leave usually only to emigrate or when they quit working. The girls leave only for marriage, in which the wife seldom works (though this is gradually changing).

The average rate of turnover for the entire plant is 2 percent per month throughout the year, compared to 8.0 percent in Phoenix. In June 1974, it was 1.3 percent and in July 0.8 percent.

At the level of department manager, only two have left MKL: one went to the United States and another went to a GE affiliate but returned after six months there. Some technicians in the ranks of the men have left the company contrary to the custom in the country (each is considered a part of the company family, following the Japanese tradition). In addition, unemployment is high and job-hopping is not easy, and MKL's wage for production workers is among the highest in Korea (a 30 percent wage increase had been announced by MKL some months before it was granted by governmental decree in 1974). Given the extensive programs for improving human relations in the company, the managers explain low turnover as a result of the fact that there are no better opportunities in Korea, and, of course, the men find it pleasant to be surrounded by numerous attractive, young girls.

Among the top male staff, only one of the total fifty left during the first half of 1974; he got a promotion and twice the salary at a textile company in the area of corporate planning. For engineers, the turnover has been only four in the past six years: two went to the United States, one to a semiconductor fabricating plant as a manager, and one to set up his own electronics company.

The low turnover is also a result of high morale among the workers and their supervisors. The concept that "people come first" permeates the plant, and a "PRIDE" program provides prizes for team and individual productivity increases each month. In addition, a Miss Motorola (MKL) is chosen each year based not on beauty but personality and general popularity among the workers; she represents the company at various occasions during the year.

Turnover varies, of course, among the departments of the plant, with some as low as 0.6 percent per month. In IC-plastics, the yearly rate is 45 percent; in DIP-ceramics, 36 percent; in MOS, 20 percent, in maintenance, 15 percent; and the machine shop, only 5 percent. One factor that may help keep turnover down is that there is a gentlemen's agreement among the three semiconductor companies in Korea that they will not pirate personnel from the others.

Promotions

The success of technology transfer is reflected not only in productivity and quality control but also in promotions of individuals within the plant. At MKL, promotion is always from within the company, if the right individual can be found; of course, rapid expansion requires hiring from outside, and the new expansion will open opportunities for present personnel to be up-graded.

The promotion policies and success in MKL is reflected in the progress of the top managers, all of whom are in their thirties. The IC-plastics manager started with the company in 1967 as a maintenance technician, while still in the university majoring in engineering. Three months later he became a maintenance trainer and upon completion of his degree became a maintenance instructor, then supervisor in maintenance, then chief, then chief in the facilities section, and a manufacturing manager two years ago.

The choice of younger men to start the MKL operation was a conscious one. Older men would not have learned as readily nor been as well prepared at the university level in engineering; nor would it have been easy to attract them away from their established positions, nor would they have had the kind of experience needed by MKL.

The manager of the DIP-ceramic line joined also in 1967, with the original group trained by Phoenix technicians and engineers. He was first a supervisor in the IC-plastics line and then a silicon plastics supervisor after training his replacement; then he became the supervisor for all third-shift operations for two years, during which he again trained his replacement. In 1969, he was made the production chief in silicon metal line; in 1971, the line chief in DIP-ceramics; two years later, the maintenance chief; and in 1974, the production manager for the line.

The manufacturing support manager also was brought in during 1967 and became a production supervisor; later he was made a process engineer on all of the different lines (save silicon metal); then head of maintenance in DIP-ceramic; and in 1973, given his present position.

The transistor manager was first a production supervisor and trainer for all lines (during 1968-69) being himself trained by a visiting engineer who stayed one month to teach the metal-can processes. He taught the trainers on all lines and in 1970 became the night-shift superintendent for all lines. During 1972-73, he was production chief in the transistor line and in mid-73 was made manager of the line.

The manager for plant engineering (facilities), the one manager not hired in the original group in 1967, came to MKL only in 1972. He was promoted to manager eighteen months ago, after working in the section for a year. He was educated in mechanical engineering and was a designer of industrial facilities for the Korean company that built the Motorola plant; he was hired, therefore, because of his intimate knowledge of MKL facilities.

The personnel manager who was also hired in the first 1967 group, had held a post in personnel with the Korean Army for fourteen years. At Motorola-Korea, he was trained by the international personnel manager, who visited Korea for one month and brought all the appropriate manuals. The MKL manager then rewrote-the sections on hiring, promotion, tardiness, absenteeism, social insurance, fringe benefits, compensation, and so forth. The practice of using employment/personnel manuals is foreign to Korean companies but has been well received by workers and supervisors. (By Korean law, all manuals have to be submitted to the government for approval; therefore, they tend to be vague and general, thereby permitting almost any interpretation and change by management through unwritten rules.)

DISSEMINATION OF TECHNOLOGY

Given the low rate of turnover at all levels of MKL personnel, there is little dissemination of technology out into the economy through transfer of personnel. In addition, the skills learned in production are not readily transferred into other types of operations. Only the more generalizable skills are readily transferred; in these cases there has been some little movement—as in facilities such as air conditioning or water treatment. One engineer, who has begun his own operation, is an example of considerable dissemination of technology. One manager has moved into another company for corporate planning; and a few engineers and machinists have moved into other companies (not semiconductors). One line manager stated that of the fifteen top men he has trained over the past seven years, five have left to go to other industries. The most useful experience is simply that in management skills and techniques, but that in quality control, maintenance, and support services is also quite valuable to companies in other industries. And some opportunities for higher salaries do exist in other companies—in fact, it is only by substantial salary increases that any manager or engineer can be moved, for they prefer security to income in the main.

Techniques learned in MKL are not applicable in general to life outside of the plant, unless they stem from facilities engineering or simple electronics; for example, one technician reported using his skills to moonlight in repair of radio and walkie-talkie equipment on occasions.

To suppliers, as indicated earlier in this chapter, the most significant transfer is in the realm of quality control and scheduling, plus cost reduction techniques.

CONTRIBUTIONS TO KOREA

The Korean government, in deciding whether to approve any foreign investment, weighs the following contributions:

1. Improvement of the international balance of payments of Korea;
2. Technical contribution to the economy;
3. Increase in employment of Korean labor;
4. Utilization of local raw materials in comparison with total raw material requirements;
5. International industrial development linkages.

MKL scores high on all but one of these criteria. Its entire product is exported, so that the value added by MKL plus that of all local purchases is regained in foreign exchange; its exports of 34 million equalled 9 percent of all industrial exports by Korea in 1973. It is almost wholly based on a transfer of technology from Phoenix to MKL. It now employs over 5,000 workers and will expand this considerably over the next few years. And it is linked wholly to international industry through being an intermediary producer. The only element on which it scores poorly is that of use of local raw materials; MKL imports almost every element that goes into the product. It does purchase locally all construction materials, materials for machinery and equipment repair, various facilities, and so forth from 200 local suppliers, but these are not in the direct production process, though they bulk large. Since Korea could not supply the basic materials for piece parts, the government, therefore, seems willing to trade off this element for top performance in the other criteria.

As to the technology contribution, the sharpest distinction can be drawn between the arrangements with Motorola-Korea and those attempted by other companies in Korea, some of which have bought technology from a U.S. company. This technology is now obsolete, and though it has provided employment, it cannot continue to support exports for long and will not keep up with the competition. In addition, having bought machines from U.S. suppliers, the companies must seek assistance in adaptation and repair and pay the cost of such assistance as needed. What MKL obtains is a continuing flow of production technology—not necessarily the latest, but the best for mass-production procedures.

Index

About the Authors

Dr. Behrman is Professor of International Business at the University of North Carolina Graduate School of Business Administration. He is engaged in research on the multinational enterprise and the role of foreign investment in economic development, and has testified before Congressional and UN Committees on these subjects. He has been a member of the Board on Science and Technology in International Development of the National Academy of Science, and is an advisor to the U.S. Department of State on problems of foreign investment.

Prior to his present position, Dr. Behrman was Assistant Secretary of Commerce for Domestic and International Business—a post he held during 1961-1964. In this office, he was responsible for the international trade and investment programs of the Department of Commerce as well as those related to stimulating domestic industrial growth. He had responsibility for trade with the Soviet countries and played a key role in the legislation on the Kennedy Round of tariff negotiations.

Dr. Behrman was born in Waco, Texas, and holds degrees of B.S. from Davidson College (N.C.); M.A. from the University of North Carolina, M.A. and Ph.D. in Economics from Princeton University. He has held faculty appointments at Davidson College, Washington and Lee University, and the University of Delaware.

Mr. Harvey W. Wallender III is Vice President of the Council of the Americas for research and planning. Since 1974, he has also acted as a consultant for a number of corporations and the U.S. Department of State in the area of strategic planning for international operations and transfer of technology. Mr. Wallender was formally Director of Planning and New Products for the Financial Services Group of the International Basic Economy Corporation (IBEC) and has acted as President of the Transoceanic Fishing Corporation and Director of Planning for the Food Products Division of IBEC. Mr. Wallender's undergraduate and graduate work in political science was done at the University of Texas followed by graduate research at the University of Madrid. He also holds an MBA from Columbia University where he is presently a PhD candidate in the Department of Management.